PERSPECTIVES ON EVOLUTION

PERSPECTIVES ON EVOLUTION

Edited by Roger Milkman

THE UNIVERSITY OF IOWA

SINAUER ASSOCIATES, INC. ● PUBLISHERS
SUNDERLAND, MASSACHUSETTS 01375

To Ernst Mayr
from all of us

THE COVER
The cover was designed by Joseph J. Vesely. The
background is in the Purple of The University of
Scranton.

Library of Congress Cataloging in Publication Data

Main entry under title:

Perspectives on evolution.

    Bibliography: p.
    Includes index
    1. Evolution. I. Milkman, Roger.
QH366.2.P45   575   81-21522
ISBN 0-87893-528-2   AACR2
ISBN 0-87893-529-0 (pbk.)

PERSPECTIVES ON EVOLUTION
Copyright © 1982 by Sinauer Associates Inc.

All rights reserved.

This book may not be reproduced in whole or in
part by any means without permission from the
publisher. For information, address
Sinauer Associates, Inc.
Sunderland, Mass. 01375

Printed in U.S.A.

9 8 7 6 5 4 3 2

CONTENTS

PREFACE

This book is intended to place some exciting recent developments in the context of evolutionary principles. Both the developments and the principles stand to gain.

Many readers will recognize, I hope, similarities to Francisco Ayala's *Molecular Evolution* (1976), a book I admired at once. *Perspectives on Evolution* is intended for everyone with an interest in the subject; it may serve, in particular, as a companion volume for a course in evolutionary biology. In the same spirit, it may bring this diverse and changing field into new focus for many students of longer standing.

The eleven chapters constitute a collection of perspectives on some of the major features of evolution. They fall, of course, into a natural classification!

Stebbins' introductory overview makes a compelling point: Unexceptionable uniformity is not in the nature of the patterns under study, and therefore a search for rigid rules and crystalline principles will keep us from seeing things the way they are. Variation is intrinsic to strategies, he asserts, as well as to morphs. The next two chapters deal with the two basic vistas of evolution, forward-looking adaptation, and phylogeny, which looks back into the past. What more can be said about these classic principles? Templeton shows in terms of modern population genetics the way in which adaptation must be understood as the result of a collaboration of forces, only one of which is natural selection. Selander proceeds from a quick allusion to comparative morphology, to a full consideration of the remarkable phylogenetic applications of the linear sequences found in proteins and in nucleic acids. He then notes the gentle irony in some recent evidence that the shapes of proteins may be even more conservative than their amino acid sequences.

The next four chapters address more restricted topics of great current importance. Given the central role of genes in the evolutionary mechanism, what do we know of genotypic variation *within* species? Ayala summarizes the dramatic results of a methodological revolution, in which an abundance of discrete phenotypic variants corresponding to genetic variants was revealed by the electrophoretic analysis of proteins from a vast range of living things. A greater degree of methodological certainty has come to paleontology, too, so that phenotypic discontinuities can no longer be attributed routinely to gaps in the fossil record. Gould's exposition of the theory of punctuated equilibrium gains strength from this new certainty and from a decade of lively debate. Stasis, relatively sudden change, speciation, and macro-evolution are seen as part of a general—but not invariant—relationship. In Chapter 6, I have tried to unify the description of selection by removing divisions between population- and quantitative genetics, between single-locus- and many-locus models, and between directional and stabilizing selection. In so doing, I have come to a conclusion whose time has evidently come (right or wrong), since it has been reached independently by both Motoo Kimura and Jack King: that allelic neutrality at many loci can coexist with strong stabilizing phenotypic selection. Next, the subject of speciation, which figures significantly in several of the earlier chapters, is addressed by Bush, who reviews a number of attributes of the process which have come to be taken for granted. Having concentrated on the subject during one of its most productive periods, Bush concludes that the most valuable contribution to our understanding of speciation would be some clear-cut demonstrations, some proofs, however limited. The time has come to make certain of what we know.

Chapters 8, 9, and 10 are introductions to subjects that are basic to our understanding of evolution. Evolution is not merely molecular, any more than it is merely a change in allele frequencies. But macro-molecules are integral to the evolutionary process, and a passing acquaintance with them will not do. Accordingly, Plapp's chapter deals with proteins as proteins. Their important structural and functional properties are described with an explicitness made possible by very recent advances. Thus, as we have done in the past with living organisms and their larger components, we come to understand what it is about proteins that may be adaptive and that may reveal phylogenetic relationships.

To most biologists with an interest in evolution, the basic properties of the nucleic acids are more familiar than those of proteins. Crawford has therefore been able to begin at another level. He describes the latest objects of molecular genetic analysis, as well as their significance for the phylogeny of bacteria. In this theater there are obviously elements of importance to all of evolution. In Chapter 10, Hunkapiller and other members of Lee Hood's laboratory discuss perhaps the most revolutionary development in our understanding of the evolutionary process: It is referred to as lateral (or horizontal) transfer, and it includes as one major product sets of intriguingly related genes called multigene families. It is now known that the genetic mechanism in evolution is not limited to the linear (vertical) modification of a gene over generations; instead, a gene may be made into extra copies and transferred to other places within the genome or to other organ-

viii

isms, where it is incorporated without evident distinction. Again, a working familiarity with this subject is requisite to an overall comprehension of evolution.

Finally we have a chapter that is speculative. Campbell reincarnates the alluring suggestion that some measure of influence on the evolutionary process itself has passed to the evolving organisms. No matter how exciting it is to sequence DNA, it is the nature of living things that makes it exciting, and the suspicion that there are still some remarkable properties to be found, inexpressible in present terms, is part of the joy of it all.

To end on a methodological note, complete citations to all references will be found in the Bibliography at the back of the book, with numbers in parentheses indicating the chapters in which they are cited. This arrangement is offered in place of an author index.

Roger Milkman
Iowa City
January 13, 1982

CONTRIBUTORS

FRANCISCO J. AYALA, Department of Genetics, University of California, Davis

GUY L. BUSH, Department of Zoology, Michigan State University, East Lansing

JOHN H. CAMPBELL, Department of Anatomy, School of Medicine, University of California, Los Angeles

IRVING P. CRAWFORD, Department of Microbiology, The University of Iowa, Iowa City

STEPHEN JAY GOULD, Museum of Comparative Zoology, Harvard University, Cambridge

LEROY HOOD, Division of Biology, California Institute of Technology, Pasadena

HENRY HUANG, Division of Biology, California Institute of Technology, Pasadena

TIM HUNKAPILLER, Division of Biology, California Institute of Technology, Pasadena

ROGER MILKMAN, Department of Zoology, The University of Iowa, Iowa City

BRYCE V. PLAPP, Department of Biochemistry, The University of Iowa, Iowa City

ROBERT K. SELANDER, Department of Biology, University of Rochester, Rochester

G. LEDYARD STEBBINS, Department of Genetics, University of California, Davis

ALAN R. TEMPLETON, Department of Biology, Washington University, St. Louis

ACKNOWLEDGMENTS

Chapters 1–9 and 11 are based on addresses given at a symposium, *Evolution in 1981,* sponsored by the Society for the Study of Evolution and by the American Society of Naturalists in Iowa City June 29–July 1. Chapter 10 was contributed in addition. The symposium and, in turn, this volume owe a great deal to May Brodbeck, Vice President for Academic Affairs at The University of Iowa, for encouragement and financial support. Deena Staub compiled the Bibliography; her effective use of information-processing equipment was invaluable.

Chapter 2/Alan R. Templeton
The author's research is supported in part by grant GM27021 from the National Institutes of Health.

Chapter 5/Stephen Jay Gould
"Although I have struggled with these issues for 10 years now, their material continues to strike me as conceptually very difficult. I lead myself into numerous blind alleys, and rely (more than I usually do) on friends and colleagues to help me out. I thank Elliot Sober for discussions on the nature and definition of group properties, Richard Burian for emphasizing the role of individuation in establishing hierarchical levels, Anthony Arnold and John Maynard Smith for persuading me that species selection must be defined narrowly in the context of units of selection, Norman Gilinsky for stressing the importance of differential origin, Sewall Wright for explaining so patiently his shifting balance theory to me, and, above all, Niles Eldredge for sharing all these concerns from the time we were graduate students together."

Chapter 6/Roger Milkman
"The central idea of this chapter was developed independently by Motoo Kimura, who responded generously and promptly to a letter from me by sharing the methods and conclusions that appeared later in his 1981 paper. I am grateful, too, for the advice and patient instruction of James F. Crow."

Chapter 7/Guy L. Bush
The author's research is supported in part by National Science Foundation grants DEB79-22881 and DEB80-11098.

Chapter 8/Bryce V. Plapp
The stereo pairs in Figures 3, 5, and 6 are copyright by Academic Press, Inc., London, and are used by permission.

PERSPECTIVES ON EVOLUTION

CHAPTER 1

MODAL THEMES: A NEW FRAMEWORK FOR EVOLUTIONARY SYNTHESES

G. Ledyard Stebbins

INTRODUCTION

The ultimate aim of scientific endeavor in any discipline is to obtain facts by reductionist methods and use them to synthesize broadly integrated theories that provide new insights into the world of nature. Integration can be achieved only by the use of a framework that illustrates the relationships among the observed facts that form the units or building blocks of the synthesis. The usual type of framework, one that has been highly successful in the physical sciences, consists of a series of generalizations or laws, such as the gas laws and the laws of thermodynamics, which can be expressed in precise quantitative terms and to which there are few or no exceptions.

In the life sciences, particularly in ecological and evolutionary biology, frameworks that consist of such laws have been much less successful than in the physical sciences. This is because of the much greater complexity that exists in living organisms, and in the biotic communities that they form, as compared to the physical world of atoms, molecules, and elementary forces. Complexity breeds uncertainty. Moreover, if the complexity is hierarchical, a slight alteration of the interacting system at one level of the hierarchy can have profound effects on the hierarchical system as a whole.

Living systems are hierarchical in three different ways. (1) The body of a multicellular animal or plant is a hierarchy of cells, tissues,

1

and organs. This hierarchy can be produced during development only by complex interactions among many genes, including those that are directly responsible for the protein molecules, the building blocks and working machines of the body, as well as genes that regulate the activity of other genes. (2) In all kinds of organisms, individuals are part of a complex hierarchy, consisting of populations, species, genera, and so on. (3) Every population is part of a biotic community that is organized in a hierarchical fashion. The environment of a population, which imposes upon it selective pressures favoring evolution in a particular direction, includes the organisms that it eats as food, predators that attack it, disease organisms to which it is susceptible, and often various kinds of symbionts without which it cannot occupy its particular ecological niche. The task of the evolutionist is to determine how these hierarchies arose, and what are their future prospects. To do this, we must go beyond the reductionist goal of analyzing the separate components of the hierarchies. We must study also the complex interactions that are the basis of their organization. The dynamics of interactions within each system, as well as between systems, are far more important than the nature of the component units.

An even more important difference exists between physico-chemical systems and those of evolutionary biology. While interactions in the physico-chemical world are essentially constant through time, those of the biological world have been continuously changing as organisms and biotic communities have become more complex. Generalizations that may have been valid and sufficient in a world that consisted entirely of microorganisms can only begin to explain the complexity of biotic communities dominated by such interactions as those between predators and prey, parasites and hosts, herbs and herbivores, and flowers and their pollinators. With respect to the modern biota, the only valid generalization is that any such law that a scientist might propose has numerous exceptions.

THE FRAMEWORK OF MODAL THEMES

I propose, therefore, a different framework. The facts of evolutionary biology can best be understood by their relationships to a modal theme. This theme should be designed to fit precisely the most common situations, but less common relationships can be expressed as variations of the central theme. Often the complexity is such that variations of variations are needed. The ultimate variations may differ so much from the modal theme that connections between them can be recognized only via intermediate, less complex variations, until simpler laws are eventually developed.

I admit that modal themes, such as those proposed below, cannot be expressed in quantitative mathematical terms as can the laws of chemistry and physics. This difficulty, however, is inherent in most of ecological and evolutionary biology. While some aspects of these fields can be adequately described and characterized by mathematical models, the entire system of hierarchical, ever-changing relationships is so complex (and full of uncertainty) that no adequate model can be devised to express all of the existing relationships within it.

THE MODAL THEME OF MENDELIAN HEREDITY

My first example is a restatement of the Mendelian laws of heredity. This restatement is nothing more than a resumé of the numerous exceptions to the original laws that geneticists have come to recognize during the eighty years since the original laws were rediscovered. As originally formulated by Mendel, they stated that hereditary information is transmitted by "factors" that are carried through egg and sperm in equal numbers, combined in the zygote, and segregated pairwise during the formation of new gametes in the adult organism. Early Mendelians visualized a direct relationship between single-factor differences and pairs of specific opposing characters in the adult. Factors were then believed to be specific particles, usually complex molecules. Their segregation and recombination was regarded as analogous to, and for teaching purposes was imitated by, removal of different colored marbles or beans from each of two bags and deposition of these units in a new bag.

The entire history of Mendelian genetics has consisted of discoveries that have modified these laws as stated by Mendel. Genetic linkage, crossing over, genes located in organelles rather than nuclei, pleiotropy, multiple locus or "polygenic" inheritance, multimer enzymes having unitary functions but coded by two or more genes, and epistatic interactions between genes at different loci—these are the principal variations, and variations of variations, that have complicated the genetic picture and rendered unrealistic a simple, direct interpretation of Mendelian laws. The most basic concept that emerged from them, that of the gene as a discrete particle, is now known to be false. The entire "chromosome" or "genophore" of a bacterial cell, containing hundreds of genes, has long been known to consist of a single, enormously long molecule of DNA. The same is true of the single DNA molecule that makes up the core of each chromosome found in nuclei of eukaryote organisms (Kavenoff et al., 1974). Structural genes are small segments of molecules that may lie adjacent to each other or may be separated by long stretches of noncoding DNA. Finally, the primary template or gene, which is transmitted from parent to offspring, may be different from the secondary template, or messenger RNA molecule, which codes directly for a protein. This is because different coding segments of the same gene may be separated from each other by one or several introns, which are eliminated when the final messenger template of RNA is constructed (Fedoroff, 1979). The Mendelian laws of heredity have been so strongly modified by successive discoveries that in their original form they are useful only for elementary teaching. They should be replaced by the following modal theme:

Biological heredity is transmitted by means of molecular templates, consisting of nucleic acids, that replicate with the aid of enzymes. These templates provide the code and the regulatory signals for the synthesis of specific proteins, the largest class of working molecules of the body. (All of the other phenomena of Mendelian heredity, mentioned above, are best regarded as variations of this modal theme.)

One advantage of the modal theme framework is that it reduces, and may even eliminate, confrontation and controversy. Such a controversy raged during the first two decades of Mendelism (1900–1920) over particulate vs. blending inheritance of quantitative characters like size and color pattern in mammals. Statistically minded geneticists such as Karl Pearson and Frank Weldon, having accumulated a vast quantity of convincing data in favor of blending inheritance in humans, rejected particulate inheritance as a general explanation. They regarded Mendel's examples and the laws derived from them as special cases (Provine, 1971). The champions of Mendelism, such as William Bateson and Wilhelm Johannsen, on the other hand, went so far in opposition to this idea that they regarded most of the variation recorded by Pearson as nonhereditary fluctuation due to environmental influences. The problem was resolved by the careful research of Nilsson-Ehle (1909), East (1916), and Castle (1916), who showed that inheritance can be particulate at the level of the gene but blending at the level of the phenotype. This situation prevails whenever the difference between the genetic potentialities of two sibling genotypes falls within the range of variation that each genotype normally displays. The complex developmental pathways between gene and character, and the ease with which these pathways can be slightly deflected, make blending inheritance at the level of the phenotype compatible with particulate inheritance at the level of the genotype.

THE MODAL THEME OF ADAPTIVENESS VS. NEUTRALITY OF POLYMORPHISM FOR ENZYME DIFFERENCES IN POPULATIONS

In the remainder of this chapter, I shall explore the modal theme framework with respect to four major problems that evolutionists currently face. The first of these is adaptiveness vs. neutrality with respect to isozyme or allozyme electromorphs of proteins. Selectionists and neutralists have been confronting each other over this issue for fifteen years. Many evolutionists not directly concerned with the controversy believe in a middle ground that is compatible with facts and arguments advanced by both sides. It can be expressed by the following modal theme.

Polymorphism with respect to allozymes is usually maintained because of slight adaptive differences, either between the alleles themselves or genes at closely linked loci. Stable frequencies may be maintained by a balance between opposing selection pressures, a balance that is subject to change as the environment changes.

Variations on this theme exist in the direction both toward neutrality and toward great adaptive differences. The differences between the M-N blood-group alleles in the modern environment of human populations may be neutral or nearly so. At the other end of the spectrum are sublethal alleles such as Hb^S (sickle-cell). The task of the population geneticist is not to decide between two sharply defined alternatives, adaptiveness vs. neutrality, but rather to determine the position that the relationship between two particular alleles occupies on a wide spectrum from neutrality to great adaptive difference. The problem is complicated by the fact that adaptiveness vs. neutrality is not an intrinsic property of a particular allele, or even a pair of alleles, but rather depends upon interactions between genotype and environment. The adaptiveness of an allele can be changed from positive to negative either by a shift in the external environment, as is that of the allele for sickling when malaria is eliminated as a serious threat to health, or by mutations or other alterations of the genotype of a population. All of these complexities can be codified as variations of the modal theme stated above.

This modal theme and its variations can be reconciled with both the relatively strong selectionist viewpoint maintained by Ayala (1974) and neutralist viewpoints such as that of Ohta (1974). She regards most mutations as nearly neutral or slightly deleterious and maintains that there is no clear cut, unconditional distinction between neutrality and adaptive significance. Given the known frequency with which environments have changed during the millions of years encompassed by the macroevolutionary time scale, one can hardly doubt that the relative position of many alternative alleles, which can be designated as a_1 and a_2, has changed in fitness from $w_1 > w_2$ through $w_1 = w_2$ to $w_1 < w_2$.

THE MODAL THEME OF SPECIATION

The second topic to which I shall apply the framework is the origin of species. This topic, more than any other, has sparked controversy, confrontation, and a persistent diversity of opinions among evolutionists as to the most probable mechanisms involved. The alternatives geographic vs. sympatric speciation, gradualism vs. saltation, adaptiveness vs. neutrality, fission vs. budding, and genetic or biological vs. morphological concepts of species have all been regarded too often as divisive pairs of alternatives between which no middle ground exists. The most recent symposium on speciation, held in Rome in May 1981 (Barigozzi, 1982), showed that no matter how species are defined, they can evolve by any one of several different routes. The following modal theme is designed to permit variations that cover all valid examples.

Populations reach the stage of distinct species when they become capable of evolving in a new direction, because reproductive isolation insulates them from other divergent populations with which they may come into contact. To reach this stage they must have acquired barriers of reproductive isolation that are effective no matter what the nature of the environment. Since reproductive isolation becomes effective more often via the accumulation of numerous genetic differences rather than via one or a few genetic steps (genes or chromosomal differences), species cannot evolve unless during at least part of the process populations are separated by spatial isolation sufficient to eliminate or drastically reduce gene flow. Genetic and chromosomal differences that promote reproductive isolation can be established independently of differences that promote morphological change or altered ecological adaptations.

In spite of its complexity, this theme must be supplemented by many variations. Visible morphological differences between species may be acquired either before or after reproductive isolation becomes complete. A few kinds of reproductive isolating barriers, such as polyploidy in plants and possibly chromosomal changes responsible for stasipatric speciation in animals (White, 1978), may arise so suddenly that little or no spatial isolation is necessary, provided that compatible mates are close at hand. On the other hand, populations of highly

6

vagile animals, such as birds and mammals, as well as wind-pollinated trees among plants, can acquire effective barriers of reproductive isolation only if they are separated from each other by a considerable distance for relatively long periods of time. In some variants, prezygotic barriers such as mating preferences evolve before postzygotic barriers such as chromosomal differences (Carson and Bryant, 1979), while in other evolutionary lines chromosomal differences are all important (Nevo and Cleve, 1978; White, 1978).

The role of natural selection in the origin of reproductive isolation varies greatly from one evolutionary line to another. In some examples, such as amphibia (Blair, 1974), the presence of character displacement indicates that direct selection favoring a reduction in the frequency of poorly adapted or sterile hybrids has played an important role at least in defining more sharply the boundaries between species. Such situations, however, are not common. In most animals and in higher plants generally, direct selection is less important than earlier authors believed it to be (White, 1978). Nevertheless, the frequent presence of genetic linkage, as well as the association of adaptive and neutral differences during great reduction in population size (Wright, 1940; Mayr, 1942), has made natural selection frequently an indirect agent in the origin of reproductive isolation.

While the modal course of events during speciation is most probably a process of "budding," which involves spatial isolation of one or a few individuals from a large ancestral population, according to the founder principle as set forth by Mayr (1954), new species can also arise in different ways. Speciation by "fission," which consists of the transformation of an entire species into two or more daughter species, is possible under certain circumstances. Because this course of events necessarily requires a longer period of time than speciation by budding, actual examples of speciation by fission are much harder to recognize. This kind of speciation is easiest to imagine as a result of primary morphological and ecological divergence, followed by extinction of intermediate populations. For instance, populations of the California salamander, *Ensatina californica,* are distributed in the form of an arc, extending from lowland southern California northward through the Coast Ranges to the northern part of the state, and from there southeastward at middle altitudes in the Sierra Nevada (Stebbins, 1949). The Sierran populations are closely related to those that inhabit the mountains of southern California, and they occur parapatrically with the lowland race of the same region, without intergradation. Consequently, one could speculate that if for any reason the races inhabiting the North Coast Ranges should become extinct, thus

7

eliminating the races that are intermediate between the mountain and southern lowland race, the two latter series of populations would then exist as separate species. To summarize, isolation by distance, accompanied by genetic differentiation and followed by breaking up of a continuous range into several separate areas of distribution, is a logical course of events to account for the origin of species, and it may have taken place many times.

With respect to the origin of species, the concept of a modal framework permits the evolutionist to adopt a pluralistic concept but at the same time focuses attention on the most probable course of events.

THE MODAL THEME OF POPULATION-ENVIRONMENT INTERACTION

During the past ten years, one of the issues that evolutionists have debated hotly and sometimes acrimoniously is whether evolution as a whole must be divided into two compartments: microevolution, which deals with divergence of populations and the origin of species; and macroevolution, which is intended to interpret the origin of major categories of the taxonomic hierarchy over long periods of time. My point of view, set forth at length in a recent paper (Stebbins and Ayala, 1981), is that "microevolution" and "macroevolution" should be regarded not as separate disciplines but as different approaches to a single holistic problem. Both approaches are necessary because of the enormous differences in time scales. The time span over which modern evolutionists have been able to make observations and conduct experiments does not exceed fifty years. While this time span could include potentially tens of thousands of generations of bacteria or other microorganisms, for other organisms the number of generations encompassed is much smaller: several hundred generations in a rapidly reproducing insect like *Drosophila,* up to 50 or 100 in annual plants or insects with longer generations such as *Orthoptera* and *Lepidoptera,* one or two hundred in small mammals like mice, and two to ten generations in most perennial plants and large animals. On the other hand, the minimal time span that paleontologists can recognize in epochs earlier than Recent or Quaternary is between 500,000 and 1,000,000 years, in other words, from 10,000 to 20,000 times as long as the maximum time span now available to microevolutionists. The difference between the temporal perspectives of the macroevolutionary paleontologists and of the microevolutionary population geneticist is comparable to that between naked-eye- and electron-microscopic perspectives of an organism.

Given this enormous difference in perspective, how can we bridge the gap between the two disciplines? I suggest that one way of doing this is by means of the following modal concept. Recognizable evolu-

8

tionary change in structure and function, which the paleontologist can follow by studying series of fossils, is the result of favorable responses on the part of populations to challenges posed by significantly changing environments. Responses consist of alterations in frequencies of many different (but interacting) genes.

Several deviations from or variations upon this modal theme can be imagined and have probably taken place. Some alterations in visible structure may be neutral or nearly so, having become established by stochastic events in small populations. Some critically different responses to environmental challenges have caused little change in outward appearance, giving rise to sibling species. These changes cannot be recognized by paleontologists. Moreover, some organisms, such as blue green bacteria ("algae") possess such wide ranges of tolerance to different environments that they can respond to environmental challenges by phenotypic modification, accompanied by relatively few and small changes in their genotypes. Their evolutionary stability over millions of years is the inevitable result of this kind of population-environment interaction (Schopf, 1978).

If this modal theme is the best expression of the most common situations, it has a direct and profound bearing upon the difference in perspective between the disciplines of microevolution and macroevolution. Experiments on populations of *Drosophila* (Anderson, 1973) have shown that even mild selection pressures can produce during a few hundred generations drastic changes in the quantitative characteristics of populations. This is more compatible with the microevolutionary than the macroevolutionary time scale. On the other hand, if one assumes a change in visible quantitative characters, such as size, which is gradual and continuous with respect to the macroevolutionary time scale, requiring tens of thousands of generations to be completed, one must postulate extremely low selection pressures, which are incompatible with the concept of response to an environmental challenge (Lande, 1976). If the modal theme stated above is applied to the macroevolutionary time scale, it leads to the conclusion that evolutionary change has not been gradual and continuous but has consisted of alternations between quantum bursts and long periods of stasis. This is consistent with the punctuated equilibrium model of Eldredge and Gould (1972), whose specific details are reviewed in Chapter 5. This opinion has already been expressed by Stanley (1979).

On the other hand, the difference in perspectives is so great that "sudden" in the language of the paleontological macroevolutionist, meaning change in less than 500,000 years, can nevertheless be gradual according to the perspective of the population geneticist. Conse-

9

quently, the data of the macroevolutionist are too crude to permit a decision as to whether or not mutations having large effects at the phenotypic level have played a role in the origin of a particular difference. The experience of experimenters in population genetics has supported the dictum of Fisher (1930) that mutations having large ultimate effects on the adaptiveness of the phenotype, even those that have relatively mild effects upon early embryogeny, are so likely to disturb adaptive harmony resulting from complex interactions between genes, that they rarely if ever convey an adaptive advantage sufficient to enable them to be established in natural populations. No data have been or can be obtained from macroevolutionary observations alone that can refute conclusions about adaptiveness and selection derived from experimental research in population genetics. "Gradual" in the language of the microevolutionist can mean "sudden" in the language of the macroevolutionist.

Elsewhere (Stebbins, 1982), I have presented an example that is supported by experimental, ecological, geographical, and paleontological evidence that illustrates well the argument of the preceding paragraph. The subgenus *Cerastes* of the genus *Ceanothus* consists of shrubs that inhabit arid and semiarid regions of the western United States and adjacent Mexico. The approximately 20 species that taxonomists recognize, several of which exist sympatrically in nature without intergradation, can all be easily crossed to produce vigorous, fertile hybrids in the F_1 and (where investigated) also the F_2 generation (Nobs, 1963). Morphological differentiation, therefore, is a true reflection of genetic and chromosomal divergence. No hidden genetic differences between these species can be detected. Ecological and geographical evidence suggests that most modern species of subgenus *Cerastes* found in northern California are not more than 10,000 to 20,000 years old. Nevertheless, fossil evidence indicates that some species of subgenus *Cerastes* have changed little or not at all for periods of 17 million years or even longer. Genetical evidence, admittedly incomplete, suggests that successive establishments of genetical changes having small effects, according to the gradualistic model, have been responsible for a remarkable burst of speciation and morphological change during the past 20,000 years on the part of evolutionary lines that previously appear to have been stable for many millions of years. This example points toward a promising method of resolving differences between microevolutionary and macroevolutionary points of view. If attention is focused upon the Pleistocene and Recent epochs of geological time, and if groups of organisms with a fossil record can be subjected to experimental research on variation within and between populations, comparisons may be obtained that could build significant bridges between the two disciplines.

Another possible bridge that may be built in the future between microevolution and macroevolution is via comparative studies of development combined, whenever possible, with morphological studies of fossil sequences. In analyzing major trends of evolution in flowering plants (Stebbins, 1974, Chaps. 6, 11), I concluded that a common sequence by which genetic changes that individually have small effects can accumulate to produce a drastic difference in developmental pathways is increasing precocity of gene action. An initial difference in pathways is produced by mutations that affect only late stages of development, and so alter only slightly the developmental pattern as a whole. As the new pathway enables the population to become adapted to an increasingly divergent factor of the environment, natural selection favors modifier or regulator genes that shift the timing of action of the mutant gene complex from late to increasingly early stages of development. Two examples were analyzed. One is the union of petals to form a bell-shaped or tubular corolla. This change is brought about by the activity of an intercalary meristem, or growing region at the base of the petals, that forms the tube. If this meristem begins activity late in the development of the corolla, after petal or corolla lobe primordia are well developed, the mature corolla has a short tube. In some of the most advanced families, like the Compositae, the tube meristem begins activity almost immediately after the petal primordia are formed, so that in most species the corolla has a long tube and very short lobes. Among different families of angiosperms, an entire series of intermediate conditions exists.

The second example is the differentiation of the flower cluster, or inflorescence. The reproductive meristem, which gives rise to flowers, differs from the leaf-producing vegetative meristem with respect to an array of histological and biochemical characteristics (Gifford and Corson, 1971). During the transition from one kind of meristem to the other, an entire battery of genes, previously inactive, must be activated, while a corresponding number are suppressed. In the Magnoliaceous genus *Michelia,* which has a solitary, relatively primitive flower, the transition takes place during the early stages of differentiation of the flower itself, so that the outer floral envelope resembles the leaf in its histological structure. In the most advanced angiosperms, such as Compositae and Gramineae, the same kind of transition occurs even before the primordia for individual flowers or flower clusters have become differentiated. Consequently, in these families, not only all parts of the flower itself but also appendages (bracts) that occur below the flowers differ radically in structure from foliage leaves. Again, the angiosperms as a whole contain an entire spectrum of conditions intermediate between these extremes.

11

THE MODAL THEME OF MOSAIC EVOLUTION

One of the most challenging ideas about evolution that has been developed during the past decade is that different kinds of genes evolve at different rates, even in the same evolutionary line. This principle is well established with respect to genes that code for different kinds of proteins, such as histones, cytochrome c, globins and fibrinopeptides (Goodman, 1976a; Fitch, 1976a; Baba et al., 1981). Even more significant is the evidence obtained by Wilson and his co-workers (Wilson, 1975; Wilson et al., 1974b, 1975; Maxson and Wilson, 1979), that an entire class of regulator genes, having only the function of activating or inactivating other genes, evolve at rates different from those of the genes that they regulate. The difficulty with this hypothesis is our imperfect knowledge of the mechanisms by which genes are "turned on" (activated for transcription) or "turned off" (inactivated) as well as equal or greater ignorance of developmental regulation at levels between genes and adult characters. I can think of at least five different kinds of regulatory mechanisms that could, by genetic changes, alter adaptive structure and function, and so play a role in the response of populations to environmental challenges. These are (1) activation of gene transcription by promoter genes and its inhibition by various genetic control mechanisms to which I will refer for convenience as inhibitors; (2) relative activity of different hormones and other growth substances, such as steroids in animals and indoleacetic acid, gibberellins and cytokinins in plants, that can be altered by mutations of genes coding for enzymes involved in their synthesis; (3) alterations in permeability of membranes, which affect the movement of these substances through the animal or plant body and can be changed by mutations affecting the structure of membrane-bound proteins; (4) alterations in cell shape, induced by pressures and tensions; and (5) alterations in the frequency of mitosis and cellular proliferation.

The validity of each of these mechanisms can be documented. Promoters and inhibitors are well known in bacteria and viruses, and their activity in eukaryotic organisms is firmly established (Rodriguez, 1981). On the other hand, these direct controlling elements are themselves subject to indirect control. In many kinds of animals, steroid hormones, particularly estrogens, have been shown to activate and inhibit gene action, most probably via their effect on DNA-based promoters and inhibitors (Saunders, 1968; Davidson, 1976). Consequently, any gene that by mutation can alter the metabolism of these hormones, whether by promoting or reducing their production, or by influencing their transport through the body, will have a regulatory effect, and so could be classified as a regulatory gene. Not all hormones have such regulatory effects on genes, but several others besides steroids, such as ecdysone in insects, certainly do. The action of plant growth substance is still unclear (Cherry, 1977), but similarities to the action of steroids are great enough so that the hypothesis, often repeated, that at least some of them act on regulators of gene action, is highly plausible.

Two examples in plants are known to me of changes in regulatory mechanisms that simulate major evolutionary changes. Basile (1979) has shown that a macroevolutionary alteration in the morphogenesis of certain liverworts can be induced by application of hydroxyproline. This amino acid, by altering proline-hydroxyproline metabolism, promotes or inhibits the growth of cell walls, which in plants play an important role in determining form. More recently, analysis of a remarkable developmental transformation in a species of *Streptocarpus* (Gesneriaceae) has shown that a major trend of evolution can be reversed by exposing germinating seeds to the action of a growth substance, gibberellin (Rosenblum, 1981).

In animals, which have flexible cell walls, cell shape can be altered by the action of contractile proteins of the cytoskeleton of the cytoplasm (Oster and Alberch, 1982). This action is controlled partly by the cellular environment and presumably also by a gene-controlled capacity of the contractile molecules to react to this environment. Movements such as these appear to be the basis of fundamental events in embryogeny, such as gastrulation. Returning to plants, pressures and tensions are responsible for the direction of cell growth and the shape of coenocytic cells in lower forms such as *Nitella* (Green, 1969; Green et al., 1970, 1971). They may also play important roles in the structuring of tissues in higher plants (Stebbins, 1968). Finally, my former associates and I have analyzed the early stages in development of a number of morphological mutants in plants (Stebbins, 1968). In each example, the first obvious effect of the mutant gene is alteration of the frequency of mitoses at a specific stage of development in a particular region of the plant meristem.

Clearly, the problem of regulation of gene action during development in multicellular organisms is highly complex and cannot at present be explained by simple generalizations. Observations and experiments at various levels and in many different kinds of organisms must be carried out and compared before valid syntheses can be attempted. I believe that the following modal theme can provide a guideline for constructing hypotheses in this all-important field.

Rates of evolutionary change usually differ by factors of two to ten or more when different organs, tissues, cells, or molecules are compared in organisms belonging to the same evolutionary line. When evolutionary rates are similar, either the characters are controlled by pleiotropic effects of the same genes or they contribute to adaptive syndromes that are subject to the same or similar selection pressures.

This modal theme is important because it reverses an emphasis on constancy of evolutionary rates that, in the past, many evolutionists have accepted almost intuitively. In addition to seeking explanations for differential rates of evolution, we should also explain as well as possible why certain organs, tissues, cells, or molecules evolve at similar rates. Clearly, neither dissimilarity nor similarity can be ascribed necessarily to different rates of mutation, although these may sometimes be involved. The stability over millions of years of bodily form in horseshoe crabs and lungfishes, of leaf structure in lycopods and ferns, of cell structure in blue green bacteria, and of amino acid sequences in histone molecules must have persisted in spite of mutations, because mutant forms have been continuously rejected by normalizing selection. Similar rates of evolution on the part of different structures need not always have a basis in natural selection, but at least the possibility of its action must be explored. At the macroevolutionary level, this kind of exploration requires close collaboration among population geneticists, developmental morphologists, comparative cytologists and biochemists, and paleontologists. The ultimate syntheses will not be the products of a single mind.

CONCLUSION

The following conclusions may be reached. First, evolutionary biology is so complex that attempts in the near future to build syntheses around the framework of rigid, all-inclusive generalizations or laws will continue to be self defeating and will lead to disputes and confrontations that generate more heat than light. Second, the framework of modal themes possesses a flexibility that enables more realistic syntheses to be elaborated. Third, this framework has been operating intuitively in the evolution of concepts of Mendelian genetics. Fourth, current disputes, such as adaptiveness vs. neutrality, geographic vs. sympatric speciation, gradualism vs. punctuated equilibria, and evolution of regulatory vs. structural genes, will be resolved more successfully on the basis of modal themes than by polarization in defense of rigidly held generalizations. Finally, at the beginning of the 1980s evolutionists find themselves in one of the most exciting fields of expanding knowledge.

CHAPTER 2

ADAPTATION AND THE INTEGRATION OF EVOLUTIONARY FORCES

Alan R. Templeton

INTRODUCTION

Evolution is often subdivided into anagenesis, or evolution within a phyletic line, and cladogenesis, or the splitting of one phyletic line into two or more lines. The major feature of anagenetic evolution is adaptation: that is, the process by which a population acquires traits that tend to enhance survivorship, mating success, and/or fertility with respect to a particular environment. (I use "adaptation" in the evolutionary sense rather than in the physiological sense.) Although Darwin entitled his book *Origin of Species,* his volume dealt primarily with the origin of adaptations. Of course, the phrase "origin of species" as used by Darwin referred primarily to the phyletic transformation of a species through time (rather than cladogenetic processes), and he viewed the process of adaptation as being primarily, but not exclusively, responsible for this transformation. Considerable attention has been paid to cladogenetic origins of species during this century, and one of the more common views regards speciation as a pleiotropic side-effect of adaptation occurring independently in two populations separated by some extrinsic barrier to gene flow (usually geographical). Under both of these views of origin of species, adaptation is clearly the primary anagenetic and cladogenetic agent. This idea has recently been challenged, with many evolutionists now claiming that both speciation and the origin of major features in macroevolution have little, if anything, to do with adaptation (Lewin, 1980). The major trends of evolution under this view are attributed instead to the process of speciation (which is regarded as involving mechanisms that are qualitatively different from those occurring during evolution within a species) and higher level processes such as "species selection" (Stanley, 1979).

In order to evaluate these claims, the mechanisms underlying adaptation and their implications must be examined very carefully. I feel the rejection of the importance of adaptation in speciation and macroevolution has often been based on an overly simplistic view of the adaptive process. I also firmly believe that speciation can under some circumstances be effectively decoupled from adaptation, but the implications of these circumstances are such that they represent a most unlikely explanation for most macroevolutionary trends (Templeton, 1980a). Consequently, I will argue that the interment of adaptation is premature and that adaptation is still alive and well and playing a critical role in the origin of species in both the anagenetic and cladogenetic senses.

SOME RARELY TOLD TALES OF SICKLE-CELL

I begin my argument by examining the mechanisms underlying adaptive evolution. To aid this examination, I will use an example—sickle-cell anemia. Although sickle-cell has become a standard example of adaptation, there are aspects of this system that are rarely told but which nonetheless offer critical insight into the mechanisms of adaptive processes.

Sickle-cell refers to an allele at the β-chain hemoglobin locus in man. This allele confers resistance to malaria when either homozygous or heterozygous (with the predominant "A" allele) and hence can greatly reduce preadult mortality (most deaths due to malaria occur in infants or children) in the malarial regions of the world. Unfortunately, another pleiotropic consequence of this allele when homozygous is a severe hemolytic anemia, known as sickle-cell anemia, which kills most homozygotes before adulthood. These effects of the *sickle-cell* allele (S allele) are reflected in the relative viability fitness estimates given in Table 1 for a West African population (based on data from Cavalli-Sforza and Bodmer, 1971). This table also gives the viabilities of the genotypes associated with a third allele at the β-chain locus, the C allele. The C allele also provides malarial resistance, but unlike S, it is a recessive allele for malarial resistance relative to the A allele. However, homozygosity for C is not associated with a severe anemia, and perhaps such homozygosity confers a degree of malarial resistance far superior to that displayed by AS heterozygotes (the fitnesses are measured with error, but for illustrative purposes, I will

16

treat the values in Table 1 as known constants hereafter). As a result, the *CC* genotype has the highest fitness: 1.3 compared to 1.0 for *AS*, the next highest. Finally, the *C* allele when coupled with *S* does lead to a hemolytic anemia that significantly lowers viability, although not nearly as severely as sickle-cell anemia.

Before the Bantu-speaking peoples of Africa expanded into west-central Africa, malaria was not an important selective agent in their environment; but with the Bantu expansion and with the introduction of slash-and-burn agriculture an extreme malarial environment was rapidly established (Weisenfeld, 1967). Initially, these Bantu populations most likely had near fixation of the *A* allele, with *S* and *C* both being very rare. However, the fitness alterations induced by the introduction of malaria lead to evolutionary changes that greatly altered these initial allele frequencies. The following model quantifies these alterations. Letting p = frequency of A, q = frequency of S, r = frequency of C, and W_{ij} = fitness of genotype ij, and assuming a single, infinite-sized population with random mating and discrete generations, the equation for the change in the frequency of S over one generation (Δq) is:

$$\Delta q = \frac{pqW_{AS} + q^2W_{SS} + qrW_{SC}}{\overline{W}} - q = \frac{q}{\overline{W}}(a_S) \qquad (1)$$

where $\overline{W}$ is the average fitness of the population at this locus and $a_S = pW_{AS} + qW_{SS} + rW_{SC} - \overline{W}$. Note that whether S increases or

TABLE 1. Relative fitnesses of some genotypes at the β-globin locus in West-African populations of man. The fitness of the *AS* heterozygote is arbitrarily set to one. Fitness differences are primarily due to the listed conditions affecting preadult viability. The fitnesses are estimated from data given in Cavalli-Sforza and Bodmer (1971).

Genotype	Fitness	Condition
AA	0.9	Malarial susceptibility
AS	1.0	Malarial resistance
SS	0.2	Anemia
AC	0.9	Malarial susceptibility
SC	0.7	Anemia
CC	1.3	Malarial resistance

17

decreases is determined only by the sign of a_S. The biological meaning of a_S is straightforward. Given a gamete bearing an S allele in a random mating population, that S gamete will pair with an A-bearing gamete with probability p to produce an AS genotype with fitness W_{AS}. Similarly, with probability q it will be in homozygous state with fitness W_{SS}, and with probability r that S gamete will be in an SC genotype with fitness W_{SC}. Therefore, a_S represents the average fitness that an S-bearing gamete will display after fertilization, minus the average fitness of the population. More technically, a_S is the conditional expected fitness deviation given that one gamete involved in zygote formation bears the S allele. The quantity a_S allows us to perform a rather remarkable trick—with it one can assign a fitness value to a haploid gamete as a function of the fitness effects it will display in diploid genotypes. This is a very important step, for, as Darwin pointed out, the only aspects of fitness that are important in adaptation are those aspects which can be transmitted to the next generation—or in more modern terminology, only the fitness effects that can be transmitted through a haploid gamete to the next diploid generation. It is precisely these fitness effects transmissible through a gamete that are measured by a_S.

The quantity a_S also has a well-defined meaning in terms of classical quantitative genetics: it is the average excess of the S allele. This definition emphasizes the fact that fitness is a phenotype and not some special entity that must be treated as qualitatively different from other phenotypic measures. Moreover, like any other phenotype at the level of whole-organism organization, fitnesses are not inherited; rather, responses to environments are inherited. Because fitness differences can alter the genetic composition of a population through natural selection, the external environment plays a direct role in shaping the evolutionary fate of the population. Indeed, this is the very essence of adaptation, as shown very clearly by equation (1). If fitness phenotypes were determined solely by the genotype rather than being inherited environmental responses, there could be no adaptation.

A final advantage of representing the allele frequency changes in terms of average excesses is that the resulting equation is very general. Virtually all classical population-genetic equations describing natural selection can be regarded as special cases of this formulation. Thus, it is not necessary to teach 1,001 Δq equations: one is sufficient. Moreover, the average excess formulation makes it clear that any factor influencing the manner in which phenotypes at one generation are transmitted through gametes to produce phenotypes in the next generation can play an active role in the adaptive process. Natural selection obviously is an important determinant of this process as it can greatly influence the types and quantities of the gametes passed

18

on to the next generation. Moreover, in the absence of different genotypic fitnesses, there can never be nonzero average excesses of fitness phenotypes. Thus, natural selection is *necessary* for adaptive evolution, but it is *not sufficient* to define an adaptive process, since other factors can and do influence the average excess. Thinking of adaptation only in terms of natural selection can be erroneous and misleading. I will now illustrate this by returning to the sickle-cell example.

As argued earlier, the premalarial Bantu population most likely had $p \approx 1$, $q \approx 0$ and $r \approx 0$. When the malarial environment was induced, these initial conditions yield $\overline{W} \approx 0.9$. Hence, the average excess of the S allele under random mating is

$$a_S \approx (1)(1.0) - 0.9 = 0.1 \qquad (2)$$

Thus, from equation (1) the frequency of S should increase, and because equation (2) is so large this initial increase will be quite rapid. This occurs because under random mating and with S initially rare virtually all S-bearing gametes become incorporated into AS genotypes with malarial resistance. On the other hand, consider the average excess of the C allele under these same initial conditions:

$$a_C \approx (1)(0.9) - 0.9 = 0 \qquad (3)$$

Because the C allele is recessive for malarial resistance, virtually all C-bearing gametes are incorporated into malarial susceptible genotypes when C is rare. Actually, the average excess of C is slightly positive when no approximations are used, but the average excess of S will still be many orders of magnitude greater than that of C. Therefore, the initial adaptive response to malaria is a rapid and large increase in the frequency of S, but practically no change in the frequency of C. Obviously, dominance and recessiveness are playing critical roles in the initial adaptive response.

S increases so rapidly, that soon a quasi-equilibrium is established with A and S in a balanced polymorphic state ($p \approx 0.89$, $q \approx 0.11$) but with C still extremely rare ($r \approx 0$). At this point, the average excesses of the S and C alleles become

$$a_S \approx (0.89)(1.0) + (0.11)(0.2) - 0.91 = 0$$
$$a_C \approx (0.89)(0.9) + (0.11)(0.7) - 0.91 = -0.03 \qquad (4)$$

Note that the average excess of the S allele is zero, reflecting its equilibrium situation with A. At this point, although the genotypes associated with the A and S alleles certainly display different fitnesses, A- and S-bearing gametes transmit identical fitness effects to the next generation. Because of this, natural selection now operates

19

as a force preventing evolution rather than causing it. On the other hand, the increase in frequency of S has changed the average excess of the C allele from 0 in equation (3) to a negative value in equation (4). This occurs because now C alleles are frequently incorporated into SC heterozygotes which have a very low fitness due to anemia. Since the C allele is still rare, its beneficial homozygous effects are virtually irrelevant. The net result is that now natural selection is operating to eliminate the C allele. Thus, starting with the initial conditions described earlier, the adaptive outcome to a malarial environment is to evolve a balanced polymorphism of the A and S alleles and to eliminate the C allele.

Recall that the C allele is associated with the genotype with highest fitness in this environment; yet natural selection insures that the fittest genotype is eliminated in this case. So much for the phrase "survival of the fittest." Moreover, note that there are basically two adaptive options to malaria: (1) become polymorphic for A and S, in which case only about 20 percent of the population is protected from malaria with the remaining 80 percent either susceptible to malaria or afflicted by an extremely deleterious genetic disease; or (2) become homozygous for C, in which case all of the population enjoys a malarial resistance greatly superior to that of AS heterozygotes and without suffering from a genetic disease. Option (1) increases average fitness from 0.9 to 0.91; option (2) from 0.9 to 1.3. I am sure most people would regard option (2) as the optimal one; yet, this is precisely the option that is eliminated during adaptation. So much for equating adaptation to "optimization."

The elimination of the C allele illustrates very well the fact that adaptation cannot be understood solely in terms of individual phenotypes. We must never forget that *adaptive processes are manifest only at the level of a population reproducing through time,* and individual fitnesses are only one factor of many influencing these processes. This point can be made more clearly by considering a hypothetical inbred Bantu population.

I will now assume a nonrandom mating population with an inbreeding coefficient of $f = 0.05$. All other assumptions will be retained. Thus, at the individual level, nothing has been altered. Under inbreeding, an S-bearing gamete will join with another S-bearing gamete identical-by-descent with probability f. Even if not identical-by-descent (an event of probability $1-f$), an S-bearing gamete can still pair with another S allele with probability q. Any heterozygotes involving S obviously must be from the noninbred portion $(1-f)$ of the population. Hence, under these rules of pairing,

$$a_S = (1-f)pW_{AS} + [f+(1-f)q]W_{SS} + (1-f)rW_{SC} - \overline{W}$$
$$a_C = (1-f)pW_{AC} + (1-f)qW_{SC} + [f+(1-f)r]W_{CC} - \overline{W}$$
(5)

20

Under the initial conditions of $p \approx 1$, $q \approx 0$, $r \approx 0$ and $\overline{W} \approx .09$ with $f = 0.05$, equations (5) become

$$a_S \approx (0.95)(1)(1.0) + (0.05)(0.2) - 0.9 = 0.06$$
$$a_C \approx (0.95)(1)(0.9) + (0.05)(1.3) - 0.9 = 0.02 \tag{6}$$

In this case the initial adaptive response is to increase the frequencies of both the S and C alleles. As C increases in frequency, more and more weight is placed upon its highly advantageous homozygous effects, but as S increases more and more weight is placed upon its highly deleterious homozygous effects. Hence, initially a_C increases but a_S decreases with time. Moreover, the increased frequency of C soon pushes the value of $\overline{W}$ beyond the maximum obtainable by an A-S polymorphism, causing the A and S alleles to acquire negative average excesses. Thus, in this case, natural selection results in the fixation of the C allele and the elimination of the A and S alleles.

Note that in both the inbreeding and random-mating examples, the initial gene pools were identical, the environmental alteration was identical, and the individual genotypic fitness responses were identical, but the adaptive outcomes were totally different. Obviously, the course of adaptation cannot be predicted from even a total knowledge of how individual genotypes respond to environments to produce fitness phenotypes. This result is not surprising in light of the average excess formulation, because system of mating is patently a critical determinant of the types and frequencies of the various genotypes into which a particular gamete type becomes incorporated. Thus, the course of adaptation cannot be explained in terms of natural selection alone.

Another way of expressing this idea is in terms of Wright's (1932) adaptive topography concept. A plot of $\overline{W}$ for all possible values of p, q and r would define a topography with two adaptive peaks under random mating: one of height 0.91 at $p = 0.89$, $q = 0.11$ and $r = 0$, and a second of height 1.3 at $p = q = 0$ and $r = 1$. The contours of the landscape are such that natural selection insures that any population starting at a point near $p = 1$ climbs the lower peak. Under inbreeding, the lower peak is altered to a height of 0.907 and a location of $p = 0.93$, $q = 0.07$ and $r = 0$, but more importantly the contours near $p = 1$ are also altered so as to insure that natural selection now causes the population to climb the higher peak. Thus, the adaptive topography is a function not only of the genotypes' fitness responses to an environment, but also of the population's system of mating. An alteration of the system of mating can potentially modify an adaptive topography as much as an alteration of the external environment. This is an important conclusion that I will use later.

As mentioned before, system of mating is not the only factor besides natural selection that is an active determinant of adaptive processes; any factor influencing phenotypic transmission through gametes is also part of the adaptive process. I will now turn to some of these other factors, starting with genetic architecture and following with the joint interactions of genetic drift and gene flow.

GENE AND SUPERGENE

An important determinant of what aspects of the phenotype are transmissible through a gamete is the genetic architecture underlying the phenotype. The number of loci, their linkage relationships, pattern of epistasis, and so forth are not just trivial complications of adaptive evolution—they are often critical determinants. A good example of the importance of genetic architecture is provided by polymorphic mimicry systems in several species of butterflies (Charlesworth and Charlesworth, 1975). In these mimicry systems, it is commonly observed that the genetic basis underlying the phenotype is a group of several alleles at different but closely linked loci that interact with one another to produce the mimetic phenotype. Such gene complexes are known as supergenes.

Charlesworth and Charlesworth have considered how such complexes could have evolved. The process begins with a major locus causing a noticeable but imperfect degree of mimicry. Suppose, for example, such a mutation causes a change in ground color from black to orange. Even though imperfect mimicry results, natural selection will often favor the increase of such an allele. Now consider, for example, a mutation at a second locus that causes a row of white spots to appear on the forewing. Further assume that if these spots appear upon a black wing they merely make the organism more conspicuous, but if they appear on an orange wing they make the mimicry more perfect. Thus, on the black background this allele is deleterious, on the orange it is beneficial. Charlesworth and Charlesworth (1975) showed that the necessary conditions for increase of such an allele when rare depend critically upon the recombination frequency with the first locus: the tighter the linkage, the more likely the second allele will be favored by natural selection. If linkage is too loose, the allele at the second locus is eliminated by natural selection, despite the fact that it increases the perfection of mimicry. Needless to say, the second allele must arise in (or recombine into) coupling with the allele for orange color.

It is also commonly observed that mimetic phenotypes in different species have totally different biochemical or developmental bases. Thus, consider the situation of a major mimicry locus with modifier mutations occurring at loci scattered throughout the genome. Each of these mutations will be associated with its own set of pleiotropic consequences, some potentially deleterious. Suppose a mutation at a modifier locus closely linked to the major mimicry locus occurs, and a mutation at an unlinked modifier locus also occurs that results in the same mimetic phenotype. Suppose further the linked locus is associated with some deleterious physiological effects whereas the unlinked locus is not. Nevertheless, the model of Charlesworth and Charlesworth (1975) predicts that the modifier at the linked locus may have a much greater chance of being incorporated into the adaptation than the more "optimal" unlinked modifier.

Charlesworth and Charlesworth (1975) therefore concluded that supergenes evolve because "the only modifier mutations which can escape elimination are ones which occur at loci which are fairly closely linked to the mimicry locus. This would obviously lead to a condition in which one region of a particular chromosome, around the original mimicry gene, contained all the loci affecting the mimetic pattern." Moreover, they show that the scheme in which loosely linked modifiers are put together by translocations and recombination modification is extremely unlikely. If appropriately placed loci did not exist a priori, such an adaptation would be impossible. Thus, these adaptive mimicry complexes can only evolve *because* of the genetic architecture.

The genetic architecture also interacts with the intensity of selection in determining adaptive outcomes. Most mutations with major phenotypic effects have many pleiotropic effects that are primarily deleterious. Indeed, sickle-cell is a good example of this. However, if selection upon one particular pleiotropic effect is strong enough, it may be sufficient to cause the allele frequency to increase, despite the deleterious nature of the other pleiotropic effects. Once again, the increase of the S allele because of its malarial resistance is a good example of this phenomenon. Once such a major allele has increased in frequency, selection will favor modifiers that suppress or circumvent the deleterious pleiotropic effects. Sickle-cell provides an excellent example of this phenomenon as well.

Malaria probably became an important selective agent in Central Africa only about 1,500 to 2,000 years ago (Weisenfeld, 1967). However, malaria has probably been selectively important in parts of the Middle East and India for considerably longer. The interesting fact is that the S allele is found in high frequency in these areas of the world as well; it is not, as commonly thought, an allele confined to black populations of African origin. However, the S allele has certainly attracted more attention in these black populations than in the Caucasian populations because the deleterious effects associated with sic-

23

kle-cell anemia have been either eliminated or greatly reduced in clinical severity, apparently due to the action of alleles at other loci in these Caucasian populations (Pembrey et al., 1980; Roth et al., 1980). Similarly, that other classic example of adaptive evolution—industrial melanism—also falls into this same pattern: the initial adaptive response is through a major locus whose expression is subsequently modified by the action of other loci (Kettlewell, 1973). Additional examples of this pattern are discussed in Templeton (1982).

The course of adaptive evolution suggested by the above examples has been confirmed by computer simulations performed by Drs. Edward Spitznagel and Theodore Reich at Washington University (personal communication). In their simulations, the initial response to selection was almost solely at the major locus which underwent rapid alteration of allele frequency until it reached a quasi-equilibrium state. Subsequent evolution of the system was more gradual and less dramatic and occurred primarily through the minor modifier loci.

One major locus has a considerable dynamic advantage over many minor loci affecting the same phenotype in the face of intense selection. The average excess at any particular locus will be quite small if the selected trait is determined by many loci, each with small additive effects, even with strong selection. Hence, the allele frequencies will change very slowly. However, with a major locus, the fitness effects are channeled primarily through a single gamete type, and hence the average excess can be quite large even in the face of deleterious pleiotropic side-effects. Hence, under intense selection, small additive polygenes often lose out to major genes as the primary cause of adaptive change (given, of course, that such genetic alternatives exist), even though the major genes often bring with them deleterious side-effects which must be eliminated or circumvented by the subsequent accumulation of modifiers. However, if selection is not intense, it is unlikely that a single pleiotropic trait of the major locus could overcome all the deleterious pleiotropic traits. In this case, the polygene architecture would most likely evolve. Note that in both of these cases the genetic architecture underlying the adaptation arises, not because it is the only genetic basis allowable for the adaptive trait, but because it is the genetic basis that interacts with natural selection to allow that particular course of adaptation. Hence, genetic architectures can play an active, not passive, role in adaptive evolution.

THE PEAKS AND PITS OF ADAPTATION

Genetic drift and gene flow are two other evolutionary forces that can have a major impact on the types and frequencies of genotypic combinations that a particular gamete-type enters into. The role of gene flow and genetic drift as agents in adaptive evolution—not neutral

24

evolution—was emphasized most strongly and effectively by Wright (1932) in his shifting balance theory. Since many misconceptions exist concerning this theory, I will now briefly discuss it.

Wright's knowledge of developmental genetics convinced him there are generally many ways of adapting to an environment, although pleiotropy insures that they are rarely equivalent in a fitness sense (recall the two unequal fitness peaks for malarial adaptation in man). Wright illustrates this conclusion with the adaptive landscape concept with its adaptive peaks of unequal height separated by adaptive valleys. His theory was primarily concerned with how balanced genetic systems could effect a shift from one peak to another in light of the fact that such peak shifts would be opposed by natural selection. However, the phrase "shifting balance" can also be used to describe the manner in which basic evolutionary forces—natural selection, genetic drift, and gene flow—interact with one another upon an adaptive landscape. To illustrate this, I will draw an analogy that is in some ways the inverse (quite literally) of Wright's original analogy of the adaptive landscape, but one for which a physical model can be easily constructed as an aid to understanding. [M. J. Kottler has brought to my attention the fact that Wright (1960) also used a similar, but not identical, analogy to the one I give below.]

Consider turning an adaptive landscape upside down, thereby transforming peaks into pits and valleys into ridges. Now let a ball correspond to a deme (*not* an individual or locus). When this ball (deme) is placed upon the inverted adaptive landscape and released, gravity will cause it to roll down to the bottom of the nearest pit, not the deepest, just as natural selection causes the demes to climb the nearest adaptive peak, not the highest. (Recall the example of sickle-cell.) Now, put some lateral motion into the balls by randomly shaking the inverted landscape. This causes the balls to roll around, even up the sides of the pits against the force of gravity, just as random genetic drift causes demes to move around the adaptive landscape, even in directions opposed by natural selection. The intensity of the shaking corresponds to the strength of genetic drift; that is, the more the inverted landscape is shaken, the smaller the deme sizes in Wright's model. During this shaking process, some balls will actually roll up the side of a pit and over a ridge, at which point gravity once again causes the balls to roll to the bottom of a new pit. This corresponds to an adaptive peak change.

As the shaking process continues, the balls preferentially come to be located in the deeper pits. The reason for this is very straightforward—it is harder to roll out of a deep pit than a shallow pit; hence,

25

as the balls roll around the inverted landscape, the ones in shallow pits are very likely to continue rolling into different pits, but the ones landing in deep pits are unlikely to make any further transitions. Another way of saying this is that there is a shift in the balance between the relative importance of gravity and random shaking in determining the movement of the balls, depending upon whether the balls are in a shallow or a deep pit. Similarly, there is a shift in the balance between natural selection and genetic drift as a deme makes transitions from peaks of unequal height and steepness. The result of this shifting balance is that populations go preferentially from lower to higher peaks through time.

Many people portray Wright's theory as if genetic drift induces peak transitions at random, and uniformly through time. However, what Wright realized is that genetic drift plus natural selection would *consistently* cause evolution from low to high adaptive peaks. Hence, adaptive evolution is far more efficient when natural selection is *not* in sole control, even though, paradoxically, natural selection is the only force actually necessary for adaptation. Similarly, gravity plus random shaking is a far more effective procedure than just gravity alone in getting all the balls to the bottoms of deep pits in my inverted landscape; yet gravity is the only force that actually causes the balls to roll to the bottom. It took the genius of a mind like Wright's to come up with such a simple yet subtle insight into the nature of adaptive evolution.

Finally, note that in shaking such an inverted landscape, pit changes (adaptive peak transitions) are very common at the beginning of the process, but as more and more of the balls come to lie in deep pits such transitions become less likely, often until there are no pit changes at all, even though the amount of random shaking may be constant throughout the entire process. This, of course, is also a result of the progressive shift in the favor of gravity in the balance between gravity and shaking. Similarly, in Wright's theory, peak shifts become less likely as the shifting balance process operates through time even if the deme sizes remain constant.

Moreover, there are further attributes of the shifting balance model that accentuate this tendency toward stasis that are not readily modeled by balls rolling in an inverted landscape. Demes are on many peaks during the initial phases of the process; hence, gene flow between them frequently acts as a random perturbing factor aiding genetic drift. But as the shifting balance process proceeds and more and more demes end up on the highest peak or a small set of high peaks, gene flow becomes more and more of a deterministic force attracting demes to the highest peaks. As more demes are brought to a single peak by the action of this type of gene flow, the more gene flow between demes acts as a factor in maintaining all the demes on that single peak, particularly if the demes on the high peaks produce most of the emigrants (Wright's "interdemic selection"). In other words, gene flow becomes more and more of a static force reinforcing natural selection and less and less of a perturbing force reinforcing genetic drift as the shifting balance process operates through time. This, of course, accentuates the shift in the balance between selection and drift, and thereby accentuates the tendency to go from dynamism to stasis. Wright (1932) has also argued that as the demes move onto the higher adaptive peaks their population sizes might also tend to increase, thereby decreasing the importance of genetic drift.

Thus, the shifting balance theory predicts that periods of evolutionary transition will be intense but brief and lead directly to a very static adaptive situation. This stasis will only be broken if the environment (and thus the landscape) is altered, if the system of mating is altered (which, as previously pointed out, also determines the topology of the adaptive landscape), if the population structure is altered to shift the balance back toward genetic drift or disruptive gene flow, or if new genetic variability occurs that adds a new, unexplored dimension to the adaptive landscape. The shifting balance theory therefore explains why adaptive evolution is rapid and why it is static; both are caused by the same underlying mechanism. There has been a tendency to emphasize the dynamic aspect in most accounts of the theory, including Wright's, but the progressive shift toward stasis inherent in the shifting balance theory is equally important in making sense out of the relationship of adaptation to speciation and macroevolutionary patterns.

THE ORIGIN OF SPECIES—ADAPTATION DEFENDED

I will now return to the original problem outlined at the beginning of this chapter: the challenge to the position that adaptation is the primary determinant of speciation and macroevolutionary trends. In essence, the basic challenge rests upon the observation of "punctuated equilibrium" (Eldredge and Gould, 1972); that is, the observation that many species apparently remain morphologically static for long periods of time, with most evolutionary changes occurring during relatively brief periods of time. It is critical to realize that punctuated equilibrium is merely a description, not a mechanism or a process. Is this description of evolution incompatible with the implications of the adaptive process? Although some recent evolutionists have answered this question in the affirmative (Lewin, 1980; Stanley, 1979), Darwin himself predicted just such a pattern of macroevolution repeatedly

27

throughout his *Origin of Species* (Templeton and Giddings, 1981). It has been argued that these statements of Darwin represent a mere handful of sentences and are not really important (Gould, personal communication, and Chapter 5.) However, the bulk of the *Origin* was devoted to the proximate causes and implications of adaptation, and Darwin did not devote much space to making explicit macroevolutionary predictions. Nevertheless, whenever he did, Darwin clearly stated that rates of evolution are unequal over long periods of time and that periods of adaptive transition are short in comparison to periods of adaptive stasis. Admittedly, these represent only a handful of explicit statements, but there are *no* statements by Darwin in which he predicts that adaptive processes should occur uniformly and continuously over long periods of geological time. Thus, the description of punctuated equilibrium is certainly compatible with Darwin's view of adaptive evolution. It is also compatible with the view of adaptive evolution outlined in this chapter.

To illustrate this point, consider the classic model of speciation in which an ancestral population is subdivided into two or more isolated subpopulations by some extrinsic barrier to gene flow. Under this model, intrinsic isolating barriers then arise as a pleiotropic consequence of the adaptive processes occurring separately within the subpopulations. However, most extrinsic geographical or ecological barriers to gene flow are temporary in nature, so the chances for speciation under this model are often a function of how rapidly adaptive divergence occurs. Unless intrinsic isolation arises rapidly enough, the extrinsically isolated subpopulations will simply fuse together upon secondary contact.

I have discussed more fully elsewhere (Templeton, 1982) the factors influencing the speed of adaptive divergence; here, I will only mention two. Adaptive divergence will be most rapid if one or both of the extrinsically isolated subpopulations is subjected to novel and intense selective pressures—a proposition with considerable empirical support (Templeton, 1981). New and intense selective pressures not only make speciation more likely in the quantitative sense that they simply speed up the rate of adaptive change, but also in a qualitative sense because of the types of genetic architecture that emerge under intense selection.

Recall that intense selection favors major genes that often have many deleterious pleiotropic effects that induce secondary adaptive processes at modifier loci. On the other hand, a less intense selective regime favors polygenes of small phenotypic effect. Hence, for the same degree of phenotypic response on a selected trait, a regime of intense selection not only achieves the response more rapidly than the less intense regime, but it also induces more major pleiotropic alterations while achieving this response. Moreover, intense selection on

28

a major locus creates conditions optimal for hitch-hiking effects. A potential example of this is provided by malarial adaptation on the island of Sardinia. In this case, intense selection favoring the X-linked allele causing glucose-6-phosphate dehydrogenase deficiency (another malarial adaptation) might have caused a hitch-hiking effect at the closely linked locus causing Protan color-blindness (Filippi et al., 1977). Thus, with intense selection, many more direct and indirect phenotypic alterations are expected on traits other than those being selected as compared to more moderate selective regimes. Hence, intense selection is far more likely to result in pleiotropic isolating barriers, and thereby speciation, because of both quantitative and qualitative considerations.

Recall also the dynamic response to intense selection: the bulk of the adaptive change occurs extremely rapidly, followed by a long period of near stasis with only minor adaptive adjustments. Moreover, the extrinsic factors that split a population in the first place are often the very same factors that induce environmental changes (e.g., see Axelrod, 1981, on the role of climatic shifts upon plant speciation in California). Therefore, the bulk of adaptive divergence under this scheme occurs shortly after the extrinsic split, followed by relative stasis. This, of course, fits the descriptive pattern of punctuated equilibrium very well.

A second determinant of the speed of adaptation is population structure. Wright (1932) predicted that adaptive evolution would be most rapid when the population is subdivided into small demes with restricted gene flow between demes. Such a structure allows his shifting balance process to operate. However, as I argued earlier, most species displaying the appropriate population structure for shifting balance would still be characterized by stasis and not continual peak transitions. Therefore, for rapid adaptive divergence to occur and lead to speciation, one or more of the extrinsically isolated subpopulations must undergo the transition from the static phase of shifting balance to the dynamic. Such a transition is quite likely for many reasons.

First, as mentioned earlier, the factors that split the population are often associated with environmental changes, and hence will induce an altered adaptive topography. Moreover, such environmental changes often induce a reduction in numbers and density of the affected species. Density changes can directly alter the system of mating, and, moreover, the nature of the relationship between density and system of mating is often nonlinear (Anderson, 1980), so that drastic alterations can sometimes occur even with a relatively modest change in density. As illustrated with sickle-cell, alterations in the system of

29

mating also alter the adaptive topography. In addition, density changes will often directly affect gene flow patterns and deme sizes, and thus could also aid in the transition to dynamic shifting balance by altering the balance between selection, drift, and gene flow.

Finally, the very act of splitting the ancestral species via extrinsic barriers obviously alters gene flow patterns. The implications of such a split is particularly important when one realizes that population structure is rarely homogeneous throughout the ancestral species. For example, Wright (1978) discusses studies on the Mojave Desert plant, *Linanthus parryae,* which indicate the neighborhood sizes in the center of the species range are of the order of 100, but in the periphery they are of the order of 10. In general, ecologically peripheral populations will often show lower deme sizes and more extinction and recolonization than central populations. As a result, the central populations should be very static under the shifting balance theory, whereas the marginal populations have far more potential to enter the dynamic phase of shifting balance. Nevertheless, this potential is often not realized because gene flow from the more numerous central populations represents a strong conservative force that can maintain the static phase of shifting balance even in the marginal areas and that can interfere with local adaptation. However, this static situation can be radically altered if gene flow from the central populations is severed, and it would also seem reasonable to assume that most geographical and ecological barriers that split a species would preferentially occur between marginal and central populations. Once gene flow from central to marginal areas has been severed, the marginal (but not the central) population could enter into the dynamic phase of shifting balance, particularly since the severing of gene flow also allows more effective local adaptation and thus the exploration of a somewhat novel adaptive landscape. Hence, ecologically and/or geographically marginal populations will often play a critical role in the speciation process resulting in a peripatric pattern of species distributions.

Note that in all of the above cases the agents causing the extrinsic split of the ancestral species are also the very same agents that can trigger the transition from static to dynamic shifting balance in one or more of the isolated subpopulations. Such a triggering should then lead to a phase of very rapid adaptive transition shortly after the split that quickly evolves into stasis. By concentrating most of the adaptive changes into a short period of time following the extrinsic split, the shifting balance mode of adaptive evolution greatly increases the chances for speciation, given a temporary extrinsic barrier. Moreover, it is patent that the resulting temporal pattern of evolutionary change is once again very consistent with the descriptive pattern of punctuated equilibrium. However, speciation is a pleiotropic by-product of adaptive processes in all the speciation models I have discussed above.

30

Consequently, the inference that the pattern of punctuated equilibrium indicates that adaptive processes are unimportant in speciation or macroevolution is not valid. This inference is based upon a simplistic caricature of the adaptive process called "phyletic gradualism" (Stanley, 1979). When adaptive processes are treated in a more rigorous and detailed fashion, no incompatibility is apparent. Consequently, I agree with Darwin that adaptation has indeed played a critical role in the origin of species.

CHAPTER 3

PHYLOGENY

Robert K. Selander

INTRODUCTION

It is generally agreed that the study of phylogeny began with Lamarck, who first consistently maintained that all types of organisms
have arisen by evolution and thus form a genealogical continuum. In
1809, he drew the first phylogenetic tree, showing, among other things,
the proposed derivation of the whales and the land mammals from the
seals (Figure 1). In this chapter, I will not attempt to review the
history of phylogenetic research or to catalogue the various methods
of phylogenetic inference. Nor will I discuss the science or art of
classification (now fashionably called an "information storage and
retrieval system"), with which phylogeny has been intimately associated since Lamarck, or the well-known differences of opinion among
"phylogenetic," "phenetic," and "evolutionary" systematists (see Mayr,
1981). My excuse for not dealing with philosophies and methodologies
of classification is that the field with which I am concerned—molecular
evolution—does not have classification as its primary objective, relevant though molecular data may be to that end.

Until recently, the construction of phylogenies and the classification of taxa was, as Ernst Mayr (1969) puts it, based on the inferred
information contents of their genetic programs. From studies of phenotypic characters, systematists attempted to infer genotypes, which
have greater explanatory and predictive value. Taxa, not characters,
are classified, although we speak of both organismal phylogeny and
molecular phylogeny. For the most part these coincide, but they may
not be completely cognate when genes are transferred among unrelated lineages, by viruses or plasmids or even directly. Wilson et al.
(1977), citing Sanderson (1976) for evidence that there is not much
effective horizontal transfer of genes, minimize this possibility, but I
think the story has yet to be told.

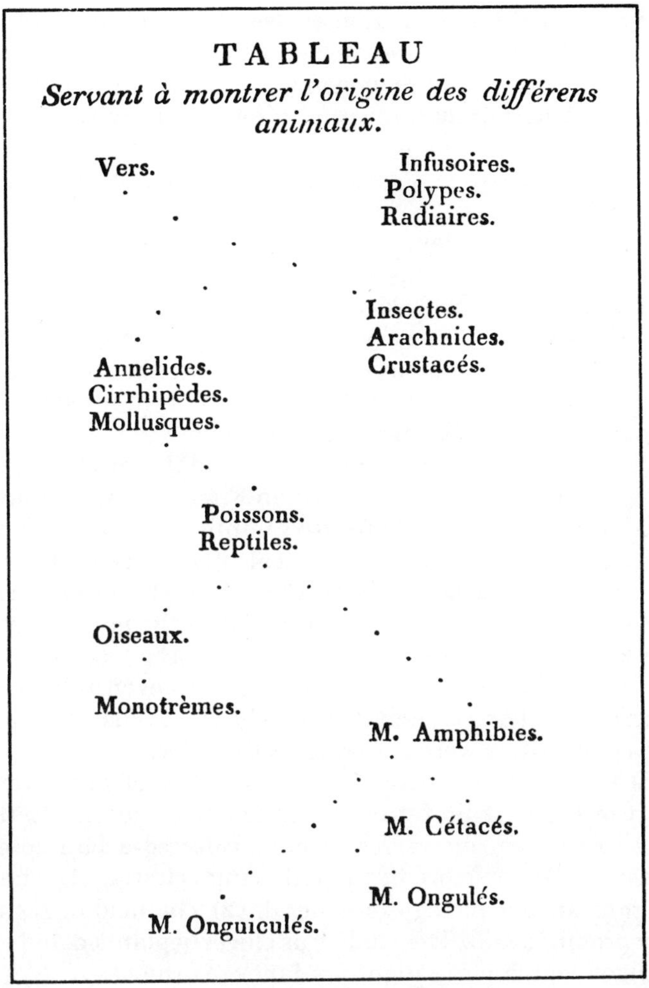

TABLEAU

Servant à montrer l'origine des différens
animaux.

Vers.

Infusoires.
Polypes.
Radiaires.

Insectes.
Arachnides.
Crustacés.

Annelides.
Cirrhipèdes.
Mollusques.

Poissons.
Reptiles.

Oiseaux.

Monotrèmes.

M. Amphibies.

M. Cétacés.

M. Ongulés.

M. Onguiculés.

FIGURE 1. A phylogeny of animals, as drawn by Lamarck in 1809.

For phylogeny, then, what we have always wanted is the structure of the genes themselves—the nucleotide sequences. The day has finally arrived. Through application of recombinant DNA techniques—including gene cutting with restriction endonucleases, splicing with ligases, and cloning in bacteria on phage or plasmid vectors—and the methodology of DNA and RNA sequencing, the molecular structures of genomes are now being directly determined. A typical announcement, appearing in 1981, was the complete sequence of the 7,433 nucleotides that make up the poliovirus (Kitamura et al., 1981).

33

The current technology, which is very new (rapid DNA sequencing was developed in 1976) and had its origins in bacterial genetics and enzymology, is being used primarily to study gene structure and regulation. Phylogenetic information of use at the organismal level has thus far been largely a by-product of this effort. But because techniques of gene cloning and sequencing are relatively simple, they will play an increasingly important role in evolutionary studies. Already there exist gene "libraries" for many species, including man, chicken, and *Drosophila*—entire genomes fragmented and cloned in *E. coli* or other bacteria.

Organisms carry evidence of their phylogenies in the nucleotides of their DNA, expressed in the amino acid sequences of proteins (which have been studied for about 25 years) and in the nucleotide sequences of their various RNAs. The record is more extensive and detailed than that provided by any combination of phenotypic characters, and much more so than the fossil record will ever be. Indeed, it extends back in time beyond the oldest fossils (Woese, 1981).

The power of macromolecules to elucidate phylogeny became clearly apparent in the mid-1960s, when Fitch and Margoliash (1967; see also Margoliash et al., 1969) used minimal mutation distance at the codon level to construct a statistical phylogenetic tree of 20 eukaryotic organisms, based solely on the information contained in the sequence of 100 or so amino acids that make up the protein cytochrome *c* (Figure 2). Here was dramatic indication of the wealth of phylogenetic information residing in macromolecules—even relatively small ones. Macromolecular structure obviously is not just "another character," as some critics were claiming at the time.

In this brief chapter, there is no possibility of my covering any major part of the vast literature of molecular evolution. Rather than attempt a comprehensive review, I have selected a half-dozen topics on the basis of one or more of the following criteria: (1) The work is highly significant for biology in general. (2) The field or technique is especially promising. (3) The study was cleverly planned and executed. And (4) the research just caught my fancy, as they say.

16S RIBOSOMAL RNA IN BACTERIA

The outstanding recent discovery in the study of phylogeny is that announced by Woese and his colleagues (Woese and Fox, 1977; Fox et al., 1980; Woese, 1981) after a decade of analyzing nucleotide sequences of 16S ribosomal RNAs in bacteria. This is a story of the phylogeny of the "prokaryotes," with important implications for the early evolution of the eukaryotes as well.

16S RNA is the medium-sized (1,540 nucleotides long) of three ribosomal RNA molecules, which, together with some 52 different

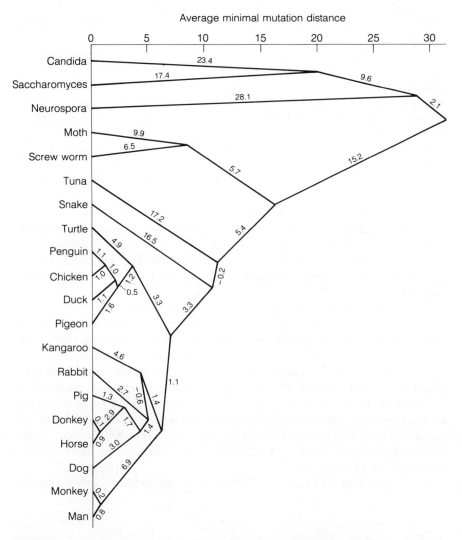

Average minimal mutation distance

FIGURE 2. Phylogenetic tree of 20 eukaryotic organisms, based on amino acid sequences of cytochrome *c*. Numbers on branches are estimated numbers of nucleotide substitutions. (After Fitch and Margoliash, 1967.)

proteins (in *E. coli*), form the ribosome. The complete nucleotide sequence of 16S RNA from *E. coli* is now available, and a model of the secondary structure was recently proposed (Woese et al., 1980*a*; Noller and Woese, 1981). Comparative studies indicate that most, if not all, conserved sequences occur in unpaired regions. The particular func-

35

tions of the various parts of the ribosome in the translation of messenger RNA remain unknown, but there is good reason to believe that 16S RNA is not merely a structural element that positions the ribosomal proteins. Thus, it is a molecular structure having a fundamental role in cellular function.

The techniques used by Woese and his associates are simple. P^{32}-labeled RNA from a given species of bacteria is cleaved with T1-ribonuclease, which cuts just to the 3' side of all guanine bases, and yields pieces ranging in size from 1 to 20 nucleotides (Figure 3). Those oligonucleotides that are longer than 6 nucleotides are then sequenced to generate a "dictionary" of "words" for comparison with similar dictionaries for other species. A typical dictionary contains 25 six-letter words; and with a word length of 6 nucleotides, a particular word is unlikely to appear by chance more than once in a given 16S molecule.

Dictionaries are now available for more than 200 species of bacteria. To compare species A and B, Woese uses a simple association coefficient (S) which expresses the proportion of nucleotides in those words occurring in both dictionaries to the number of nucleotides in all words in both dictionaries. S ranges from 1 (identity) to nearly 0 (unrelated).

Dendrograms showing relationships among various groups of bacteria (see, for example, Figure 4) are constructed by cluster analysis from a matrix of association coefficients (Table 1). Other, more complicated but perhaps more informative, methods of constructing trees have not been used because of uncertainty as to the details of relationship between S and the actual number of nucleotide differences. However, this difficulty will soon be overcome when complete nucleotide sequences become available for a variety of species. On the assumption that all lines of descent have evolved at similar rates, Woese and his colleagues (Woese et al., 1980b) believe that the dendrograms approximate genealogies.

Studies of the 16S ribosomal RNA were undertaken in an attempt

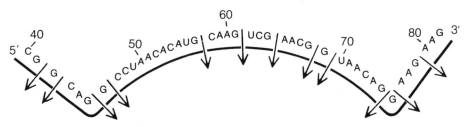

FIGURE 3. T1-ribonuclease cleaves RNA just to the 3' side of all guanine bases, yielding oligonucleotides of varied lengths. (After Woese, 1981.)

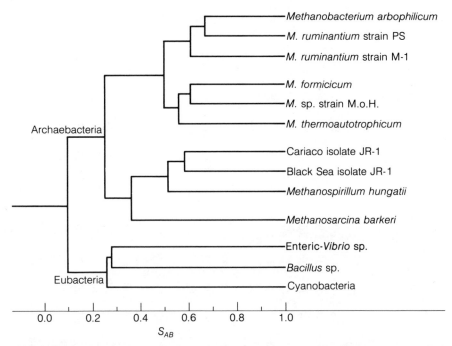

FIGURE 4. Dendrogram of relationships of methanogenic bacteria (classified as archaebacteria) and eubacteria (Enteric-*Vibrio, Bacillus* sp., and Cyanobacteria), constructed by average linkage clustering (between merged groups) from association coefficients (S_{AB}) for pairwise comparisons of various species, strains, and isolates. (After Fox et al., 1977.)

to improve bacterial systematics, which have never been in good shape because bacteria are simple organisms, with phenotypes whose information content is low. As Woese screened various types of bacteria, he found that, while in many cases his RNA genealogies were consistent with traditional phenotypic classifications, in some cases they were not. For example, it was found (Woese et al., 1980*b*) that the mycoplasmas (bacteria lacking a cell wall) are phylogenetically not so far-out as many had thought, and that they are polyphyletic. In addition, great variation was found in the "evolutionary depth" of genera—this being shallow among the genera of the Enterobacteriaceae, with $S = 0.7 - 0.8$ (a fact already known from DNA hybridization), and deep in the case of *Clostridium,* which differs from other genera by $S = 0.03 - 0.35$. This situation is comparable to that described in higher animals by Wilson and his colleagues (Wallace et al., 1971;

37

TABLE 1. Coefficients of association (S) for various organisms, based on comparison of 16S ribosomal sequences. 1–3 represent eukaryotes; 4–8, eubacteria; 9, organelle of a eukaryotic plant; and 10–16, archaebacteria.

	1	2	3	4	5	6	7	8	9	10	11	12	13	14	15	16
1 *Saccharomyces cerevisiae*	—	.29	.33	.05	.06	.08	.09	.11	.08	.11	.11	.08	.08	.10	.07	.08
2 *Lemna minor*	.29	—	.36	.10	.05	.06	.10	.09	.11	.10	.10	.13	.07	.09	.07	.09
3 *L cell*	.33	.36	—	.06	.06	.07	.07	.09	.06	.10	.10	.09	.07	.11	.06	.07
4 *Escherichia coli*	.05	.10	.06	—	.24	.25	.28	.26	.21	.11	.12	.07	.12	.07	.07	.09
5 *Chlorobium vibrioforme*	.06	.05	.06	.24	—	.22	.22	.20	.19	.06	.07	.06	.09	.07	.05	.07
6 *Bacillus firmus*	.08	.06	.07	.25	.22	—	.34	.26	.20	.11	.13	.06	.12	.10	.07	.09
7 *Corynebacterium diphtheriae*	.09	.10	.07	.28	.22	.34	—	.23	.21	.12	.12	.09	.10	.10	.06	.09
8 *Aphanocapsa*	.11	.09	.09	.26	.20	.26	.23	—	.31	.11	.11	.10	.10	.13	.10	.10
9 *Chloroplast (Lemna)*	.08	.11	.06	.21	.19	.20	.21	.31	—	.14	.12	.10	.12	.12	.06	.07
10 *Methanobacterium thermoautotrophicum*	.11	.10	.10	.11	.06	.11	.12	.11	.14	—	.51	.25	.30	.34	.17	.19
11 *Methanobrevibacter ruminantium*	.11	.10	.10	.12	.07	.13	.12	.11	.12	.51	—	.25	.24	.31	.15	.20
12 *Methanogenium cariaci*	.08	.13	.09	.07	.06	.06	.09	.10	.10	.25	.25	—	.32	.29	.13	.21
13 *Methanosarcina barkeri*	.08	.07	.07	.12	.09	.12	.10	.10	.12	.30	.24	.32	—	.28	.16	.23
14 *Halobacterium halobium*	.10	.09	.11	.07	.07	.10	.10	.13	.12	.34	.31	.29	.28	—	.19	.23
15 *Sulfolobus acidocaldarius*	.07	.07	.06	.07	.05	.07	.06	.10	.06	.17	.15	.13	.16	.19	—	.13
16 *Thermoplasma acidophilum*	.08	.09	.07	.09	.07	.09	.09	.10	.07	.19	.20	.21	.23	.23	.13	—

Wilson et al., 1974a), in which the intrageneric distinction among frogs corresponds—according to immunological distance measures—to that between higher categories of mammals (see related paper by Cherry et al., 1978).

A major surprise came when the methanogenic bacteria, which live anaerobically and generate methane by the reduction of carbon dioxide, were examined (Fox et al., 1977), for these turned out to be unlike any bacteria that had been studied (Figure 4). Later, some other aberrant forms—the extreme halophiles and thermoacidophiles—were analyzed. All three groups were somewhat alike and differed from other bacteria (Figure 5). In fact, by the evidence from 16S rRNA, the methanogens (the "deep" group) are as closely related to the eukaryotes as to the other, more conventional bacteria.

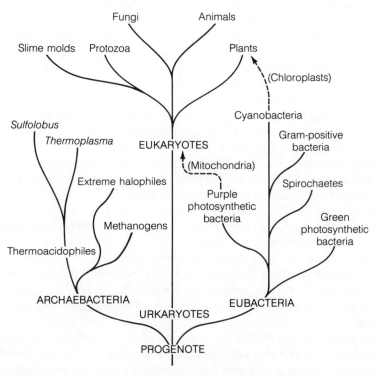

FIGURE 5. Three kingdoms of life, a phylogeny based on nucleotide sequences of 16S ribosomal RNA. (After Woese, 1981.)

Woese and his colleagues propose to call the methanogens and their relatives the *archaebacteria,* in distinction to the *eubacteria.* Believing that 16S RNA sequence divergence measures evolutionary time (not "progression"), they argue for a great antiquity of the archaebacteria—as great or greater than that for any other group of bacteria—and as the name implies, they believe that they were the dominant form in the primeval biosphere, where they lived by reducing carbon dioxide in the virtual absence of oxygen. Hence we have two prokaryotic lines—a third kingdom of life has been added (Figure 5). See Osawa and Hori (1980) for critical comments on the hypothesis proposed by Woese and his colleagues and an alternative explanation for the characteristics of the "archaebacteria," based in part on sequence data for 5S rRNAs and RNA-binding proteins.

It has long been recognized that the forms now classified as the archaebacteria are individually peculiar—but in each case the aberrancy was attributed to adaptation to some special niche or a biochemical "quirk." However, the measurements of RNA sequence divergence show that these are general characters of a new group of organisms. This observation has stimulated many other types of studies of bacteria to determine whether the RNA pattern is unique. Woese predicted that the archaebacteria would prove to be as different from the eubacteria in their overall molecular phenotype as either group is from eukaryotic cells. And there is strong evidence that this is the case (Table 2). For example, in the archaebacteria: (1) The cell-wall type varies but none has the muramic-acid-based type found in the eubacteria. (2) Membrane lipids are straight chains, as in eukaryotes, but unlike those in eubacteria. (3) The transfer RNAs are peculiar— for example, uracil in the "common arm" is not methylated to form thymine; instead it is modified to form a pseudouridine. These and other features of the archaebacteria do not seem to be related to the circumstance of their occupying unusual niches.

To Woese, the importance of this discovery lies in what it may tell us about the early history of life, since with three approximately equidistant lines of descent, rather than two, we are in a better position to reconstruct the common ancestor of all organisms and to trace the evolution of the eukaryotic cell. He proposes a bold hypothesis involving an ancestral "progenote" in which the cell and the genetic and translational mechanisms were simpler than in any extant organism.

Suppose that all of this is true, and that a distinction between the archaebacteria and the eubacteria has existed for at least 3.5 billion years. This would leave less than a billion years for the origin of life and evolution of the first bacteria. Yet the evolutionary changes that occurred within each bacterial kingdom in 3.5 billion years are small compared with the differences that separate archaebacteria from true bacteria. To account for this, Woese proposes that both bacterial king-doms and the eukaryotic stem arose from "progenotes" in which basic molecular characters (cell wall, synthesis of lipids, coenzymes, RNA polymerase subunit structure, etc.) were in the process of developing. He further proposes that translation mechanisms were rudimentary, and that many modifications in each of the three kingdoms have evolved independently to "fine tune" these processes. This theory can be further tested as we learn more about mechanisms of gene expression.

What about the eukaryotes? In Woese's scheme (Figure 5), the base of the line is labelled "urkaryote." This interpretation is based on growing evidence that the eukaryotic cell is a phylogenetic chimera; for example, the chloroplasts of plants coming from cyanobacteria-like forms and the plant mitochondrion from the purple photosynthetic bacteria. What about the nucleus: a single origin or multiple sources? To what extent are eubacterial and archaebacterial genes represented in the eukaryotic chimera? (There is a report that at least one of the eukaryotic genes encoding cytoplasmic ribosomal protein is homologous with a ribosomal gene of an archaebacterium; Matheson et al., 1980.) In all, it seems likely that studies of molecular variation in bacterial RNA will ultimately lead to major advances in our understanding of the early phylogenetic history of higher organisms.

MITOCHONDRIAL DNA SEQUENCES

In the past few years, much attention has been given to the assessment of variation among organisms in the nucleotide sequences of mito-chondrial DNA (mtDNA). The mitochondrion carries a manageable amount of DNA in comparison with the nuclear genome, which in man, for example, consists of some 7 billion nucleotides, only about 65 percent of which are in single copy sequences. A major achievement in biology, announced in 1981, was the complete sequencing of the human mitochondrial genome, consisting of 16,569 base pairs (Anderson et al., 1981). (Bovine and mouse mitochondria have also been sequenced, but the results are not yet published.) Future phylogenetic studies will be based on direct nucleotide sequencing, but to this point most comparative work has involved the indexing of sequence differences through the use of restriction endonucleases (Lansman et al., 1981). Nei and Li (1979) developed a now generally used method of estimating numbers of nucleotide substitutions between populations or species from comparisons of cleavage maps or the distributions of fragment lengths. And they defined a coefficient of nucleotide diver-

41

TABLE 2. Morphological and molecular characteristics of the three kingdoms of life.

	Archaebacteria	Eubacteria	Eukaryotes
Cell size (linear dimension)	about 1 micrometer	about 1 micrometer	about 10 micrometers
Cellular organelles	absent	absent	present
Nuclear membrane	absent	absent	present
Cell wall	variety of types; none incorporates muramic acid	variety within one type; all incorporate muramic acid	no cell wall in animal cells; variety of types in other phyla
Membrane lipids	ether-linked branched aliphatic chains	ester-linked straight aliphatic chains	ester-linked straight aliphatic chains
Transfer RNAs			
Thymine in "common" arm	absent	present in most transfer RNAs of most species	present in most transfer RNAs of all species
Dihydrouracil	absent in all but one genus	present in all species	present in most transfer RNAs of all species
Amino acid carried by initiator transfer RNA	methionine	formylmethionine	methionine

Ribosomes			
Subunit sizes	20S, 50S	30S, 50S	40S, 60S
Approximate length of 16S (18S) RNA	1,500 nucleotides	1,500 nucleotides	1,800 nucleotides
Approximate length of 23S (25–28S) RNA	2,900 nucleotides	2,900 nucleotides	3,500 nucleotides or more
Translation-elongation factor	reacts with diphtheria toxin	does not react with diphtheria toxin	reacts with diphtheria toxin
Sensitivity to chloramphenicol	insensitive	sensitive	insensitive
Sensitivity to anisomycin	sensitive	insensitive	sensitive
Sensitivity to kanamycin	insensitive	sensitive	insensitive
Messenger-RNA binding site AUCACCUCC at 3' end of 16S (18S) RNA	present	present	absent

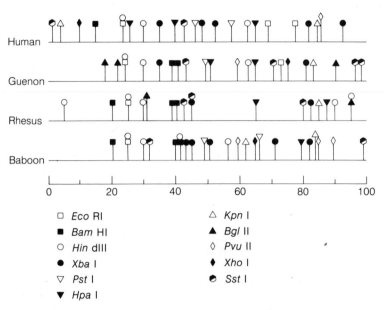

FIGURE 6. Cleavage maps from mitochondrial DNA of several primates. Symbols designate sites cleaved by various restriction endonucleases. (After Brown et al., 1979.)

sity, which is equal to the average number of nucleotide differences per site between two randomly chosen DNA sequences (see also Nei and Tajima, 1981). Similar measures have been derived by Upholt (1977) and Kaplan and Langley (1979).

Evolution in mtDNA in primates, based on cleavage maps. Using eleven restriction endonucleases, Brown et al. (1979) constructed cleavage maps for the Guinea baboon (*Papio papio*), rhesus macaque (*Macaca mullata*), green monkey or guenon (*Cercopithecus aethiops*), and human (Figure 6). The many restriction sites at which the maps differ seem to be distributed randomly throughout the genome, and the mtDNAs are estimated to differ by an average of 25 percent in nucleotide sequence, sharing only from 18 percent to 25 percent of cleavage sites in common.

Table 3 compares the extent of sequence divergence in mitochondrial DNA and single-copy nuclear DNA in several pairs of higher primates. Differences in mitochondrial DNA exceed those in nuclear DNA by fivefold on the basis of m (the minimum number of substitutions per base pair) and by tenfold on the basis of p (the estimated number of substitutions per base pair, corrected for multiple substitutions at some cleavage sites). Thus, mtDNA appears to evolve 5 to 10 times faster than single-copy nuclear DNA.

Why does mtDNA evolve so rapidly? There are several hypotheses, including the following. (1) The mutation rate is high because editing is lacking or inefficient; e.g., there apparently is no enzymatic provision for the excision and repair of thymine dimers. (2) Because the cell is essentially polyploid for mitochondrial genes, one genome mutating may have little effect on the cell or organism (a "dispensibility" factor). Whatever the reasons, mtDNA has great promise for providing information on the phylogeny of closely related organisms.

In an attempt to elucidate the phylogeny of the higher primates, Ferris et al. (1981*b*) compared humans, gorilla (*Gorilla gorilla*), two species of chimpanzees (*Pan troglodytes* and *P. paniscus*), orangutan (*Pongo pygmaeus*), and gibbon (*Hylobates lar*) with respect to cleavage maps obtained with 19 endonucleases. About 50 cleavage sites were recorded for each species, and differences between maps were evident at 121 positions. Incidental to their main purpose, they obtained information on the relative resolving power of their technique, as shown in Table 4.

Molecular data, including mtDNA, have not given a consistent indication of branching order of man-chimpanzee-gorilla. For mtDNA, the most parsimonious tree has chimpanzee and gorilla diverging after the human line split off from the African ape lineage (minimal-mutation parsimony method yielding 67 mutations at the 42 positions that provide information). (The tree is precisely like that of Simpson's (1963), based on morphology!) A tree with a three-way split (human-chimpanzee-gorilla) requires 71 mutations. It is estimated that at least 10 times more genetic information will be required to resolve the

TABLE 3. Comparison of extent of sequence divergence in mitochondrial and nuclear DNA in several primates.

| | PERCENT OF SEQUENCE DIFFERENCE | | |
| | mtDNA | | Single-copy nuclear DNA |
Species compared	Minimum	Estimated	Estimated from hybridization
Baboon-rhesus	12.7	24	1.4
Baboon-guenon	12.6	24	2.2
Guenon-rhesus	12.0	21	2.0
Human-baboon	13.7	29	4.7
Human-guenon	12.5	23	6.3

45

TABLE 4. Resolving power of molecular techniques in chimpanzee-human comparison.

Technique	Resolving power
Protein immunology and DNA hybridization	Continuous scale, error large for small differences
Electrophoresis	23 loci; 7 differences
Amino acid sequencing (9 proteins)	963 amino acid positions; 3 substitutions detected
mtRNA restriction maps (19 enzymes)	44 restriction site differences

branching order. This could come from more restriction site mapping or, easily, from direct nucleotide sequencing of mtDNA and/or nuclear DNA.

Polymorphisms in mtDNA in a racially diverse sample of humans was demonstrated by Brown (1980), using 18 restriction enzymes. Each of the 21 humans studied could be characterized by the digests; and all between-sample differences could be explained by single base substitutions. The average pairwise difference was estimated at .0036 substitutions per base pair (the diversity measure of Nei and Li, 1979) and was the same within and between races. Earlier, the rate of evolution of mammalian mtDNA was estimated to be 0.01 substitutions per base per lineage per million years (Brown et al., 1979). (Like all estimates of absolute rates, this one is based on dates of divergence supplied by paleontologists; and none of the dates is very reliable.) In any event, the conclusion is that the heterogeneity in the human species could have been generated from a mitochondrially monomorphic population beginning only 180,000 years ago. A basic similarity in the cleavage patterns of mitochondria from the major racial groups of man indicates that they have diverged since that time.

Ferris et al. (1981a) studied sequence divergence within the lowland gorilla, common chimpanzee, pygmy chimpanzee, and orangutan; a total of 86 individuals, of which 27 were studied intensively, with 25 enzymes. These sequences were compared with those of humans, with the following results: For humans, mean pairwise diversity of a racially heterogeneous sample was, as noted earlier, 0.36 percent; pygmy chimpanzee, 1.0 percent, common chimpanzee, 1.3 percent, and between species of chimpanzee, 3.7 percent; orangutan, from 0.6 to 5.5 percent, with the biggest difference between populations in Borneo and Sumatra; and gorilla, 0.6 percent (maximum = 0.9 percent), but this is an underestimation, since the mountain gorilla was not studied.

The great apes apparently have a higher level of mtDNA diversity than man. The mtDNA data indicate that the common and pygmy chimpanzees diverged 1.3 million years ago, which is consistent with protein evidence (Bruce and Ayala, 1979; Sarich, 1977), but there are no fossils with which to test this estimate. For orangutans, mtDNA points to a divergence between Borneo and Sumatra at 1.5 million years B.P., but again, no fossils are available.

In sum, the extent of intraspecific divergence in the higher primates is related to the age of the species rather than to its present population size. The great apes are relatively old as species, although they presently have rather small (N = 10,000), localized populations; while modern man (*Homo sapiens sapiens*) presently has a large population but is perhaps only 35,000 years old.

Avise et al. (1979) studied mitochondrial DNA sequence relatedness in natural populations of three species of mice of the genus *Peromyscus*. Based on the application of six endonucleases in various combinations, their data indicate a nucleotide diversity for *P. polionotus* of about 1 percent, which is similar to that of the chimpanzee. An extensive study of intra- and interspecific variation in the mitochondrial genome of two species of *Rattus* was recently reported by Brown and Simpson (1981).

DIRECT MEASUREMENT OF RATES OF NUCLEAR NUCLEOTIDE SUBSTITUTIONS

Direct comparison of homologous DNA sequences enables one to evaluate the extent of sequence divergence in various functional or structural units of genomes, such as coding regions, intervening sequences, 5'- and 3'-flanking transcribed noncoding regions, flanking nontranscribed regions, and pseudogenes (functionless genes). In addition, for coding regions, it is possible to compare the degrees of divergence caused by replacement substitutions—those leading to amino acid changes, and silent substitutions—those leading to a synonymous codon change. Most of the silent substitutions occur in the third position of codons. The data were recently reviewed by Jukes (1980) and by Miyata et al. (1980).

There is abundant evidence that large numbers of silent nucleotide substitutions occur during the evolutionary divergence of lineages. Several estimates indicate that the rate of substitution at the third position of codons is about four times greater than that at the first and second positions. And Jukes (1980) estimates that half of the nucleotide substitutions occurring during the evolutionary divergence of genes in animals, bacteria, and viruses are silent changes. Together

47

with Kimura (1977, 1980), he believes that most or all of them are selectively neutral. Several lines of support for this argument have been advanced. First, it is unlikely that silent nucleotide differences between related species are adaptive in reference to mRNA secondary structure or tRNA utilization, because this would mean that related species have marked differences in the manner of expression of each pair of homologous genes. How big are the differences? For two species of sea urchins, *Stronglyocentrotus purpuratus* and *Psammechinus miliaris,* partial or complete DNA sequences for four histone genes are available. Although there are only 2.5 amino acid replacements per 100 codons compared, there are 37 silent substitutions per 100 codons.

Second, pseudogenes are DNA segments that show homology to a functional gene but contain nucleotide changes, such as frameshifts, that prevent their expression (Proudfoot, 1980). In the globin gene family as many as one in four genes may be pseudogenes. Since they apparently are functionless, their rate of nucleotide substitution is of special interest. Complete nucleotide sequences are now available for three pseudogenes—mouse $\psi\alpha3$, human $\psi\alpha1$, and rabbit $\psi\beta2$. In a recent paper, Li et al. (1981) have estimated rates of substitution in the evolution of these pseudogenes from the corresponding functional genes. For the pseudogenes, the rates are the same for all three codon positions, being 4.6×10^{-9} substitutions/nucleotide/year. This is even higher than the value of 2.64×10^{-9} for position 3 in the functional genes. (For positions 1 and 2, the rates were 0.71 and 0.62×10^{-9}, respectively.) The pseudogene rate is comparable to that reported by Perler et al. (1980) for the silent substitution rate in the C-peptide region of the preproinsulin gene (7.0×10^{-9}). The C region apparently has no function other than holding together the polypeptides that will later form a protein.

Kimura (1977, 1980) has maintained that the rate of base substitutions in pseudogenes should approach the upper limit set by the mutation rate. The fact that the estimated rate for silent substitutions in functional genes is not very different makes it difficult to maintain that the majority of such changes are mediated by selection.

When the theory of molecular evolution by neutral substitutions was proposed by Kimura (1968*a, b*) and King and Jukes (1969), many evolutionists reacted by asserting an unwavering faith that all aspects of molecular evolution would, on thorough analysis, turn out to be mediated by the Darwinian process of natural selection. This was at the time a sensible strategy, given the history of evolutionary biology. It was even predicted that nucleotide substitutions in the third position of codons are constrained by the abundance of various nucleotides or of different tRNAs. The data already show that they are not constrained very much, but there is also sufficient evidence regarding patterns of codon usage and frequencies of transitions and transversions to indicate the substitution is not entirely random (see, for example, Perler et al., 1980).

If the position of the selectionists was difficult to maintain in the 1960s, consider the circumstance today, when Orgel and Crick (1980), Doolittle and Sapienza (1980), and others are seriously maintaining that perhaps half the eukaryotic genome is "junk" sequence whose only "function" is to replicate.

MOLECULAR CLOCKS

What can one say about the controversial and much-discussed subject of molecular clocks that has not already been said? The extensive evidence for their existence and value in evolutionary studies was summarized at length by Wilson et al. (1977), Sarich and Cronin (1976), and Carlson et al. (1978). And the major arguments against the existence of useful molecular clocks were presented in a series of papers by Goodman, Tashian, and Simons in *Molecular Anthropology,* published in 1976 (Goodman et al., 1976). Among more recent contributions to the controversy are those of Zuckerkandl (1976), Corruccini et al. (1979, 1980), Hartl and Dykhuizen (1979), Maxson and Maxson (1979), Avise et al. (1980), Efstratiadis et al. (1980), Sarich and Cronin (1980), Scanlan et al. (1980), Korey (1981), Goodman (1981*a*), Larson et al. (1981), and Wyles and Gorman (1980).

The possibility that macromolecules are evolving at rates sufficiently constant to serve as evolutionary clocks was first suggested by Zuckerkandl and Pauling (1962, 1965). Subsequently, the announcement, by Sarich and Wilson (1967*a, b*) of biochemical evidence that the human lineage branched from that leading to the African apes only about 5 million years ago, rather than 15 to 30 million years ago, as claimed by students of the fossil record, caused a major stir in paleontological circles. Claims that proto-chimpanzees and proto-gorillas existed 20 million years ago were withdrawn (Andrews, 1974; but see Simons, 1976). And the key piece in the pongid-hominid puzzle—*Ramapithecus*, a supposed hominid dated at 14 million years B.P.—was reexamined and found by some to be not clearly on the hominid line (Walker and Andrews, 1973; Greenfield, 1974). However, Simons (1976, 1977) and Simons and Pilbeam (1978) have continued to maintain that *Ramapithecus* is a hominid, based on study of additional specimens; and Walker (1976) has now come around to this view. Yet, many remain unconvinced that *Ramapithecus* provides evidence that the hominid line was distinct from the pongid line as early as 14 million years ago. The credibility of primate paleontologists has been weakened by the advent of molecular techniques of assessing relationships and estimating times of divergence. Certainly it is not comforting to be told (by Walker, 1976) that, "as records of fossil groups go, the hominid story is one of the more complete," and then to learn, in the same book, that Walker sets the *Pan/Gorilla* divergence at about 3.5 million years, while Simons is convinced that it occurred 18 million years ago. The situation may, however, be improving: In a recent paper, two prominent paleontologists suggested that "paleobiologists should undertake careful analysis of the fossil record . . . to find divergent lineages that can be documented with some confidence" (Jacobs and Pilbeam, 1980). Some would say, "It's about time." In any event, we are unlikely again to see, as we did in 1972, a Bjorn Kurtén dismiss the molecular and chromosomal evidence of the phylogeny of man and apes, maintain that only fossil evidence is capable of elucidating phylogenies, interpret the fragmentary fossil record to indicate that the ancestry of man and of the apes and monkeys has been separate for more than 35 million years, and attribute the anatomical, chromosomal, and molecular similarities between man and apes to parallel evolution.

Shown in Figure 7 is a composite evolutionary clock based on amino acid sequences for cytochrome c, myoglobin, hemoglobin α-chain, hemoglobin β-chain, two fibrinopeptides (A and B), and the insulin C-peptide in 17 species of mammals. The number of nucleotide substitutions was estimated by Langley and Fitch (1974; see also Fitch and Langley, 1976; and Fitch, 1976a), using a maximum likelihood procedure. Their analysis indicated that there is a significant deviation from uniformity in both the combined or overall rate of change and in the relative rates of change in the seven proteins studied. The variance in the molecular clock, as measured by Langley and Fitch, is about twice that expected in a strictly stochastic clock, such as that provided by radioactive decay. Notwithstanding these shortcomings, Fitch (1976a) concludes that "averaged over longer time intervals and greater numbers of proteins the estimates ought to be fairly good and, if so, to be linear with respect to paleontological time."

In the clock shown in Figure 7, the primates fall below the line drawn from the origin through the outermost point, which represents the marsupial-placental divergence, and about which most other points cluster tightly. Wilson, Sarich, and the other proponents of the clock assume that the times of divergence supplied by the paleontologists are in error—that they really were more recent than shown. But Goodman, Tashian, Kohne, and many primate paleontologists believe that there has been a reduced rate of molecular evolution in the primates, the divergence dates being correct. There is no possibility of my even summarizing all the various arguments and data sets relevant to the problem of the extent of lineage-specific variation in

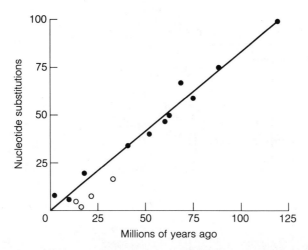

FIGURE 7. Nucleotide substitutions versus paleontological time. Nucleotide substitutions are numbers expected on basis of a maximum-likelihood fit to maximum-parsimony solution for seven proteins from 17 species of mammals. Dates of divergence (in millions of years) are based on fossil evidence. Circles indicate divergences within the primate line. (After Langley and Fitch, 1973, 1974.)

rates of molecular evolution. My impression is that the weight of evidence favors the view that molecules are in general evolving at fairly constant rates—and that times for a clock based on pooled data from several proteins must be taken seriously. The relative rate tests (Wilson et al., 1977) seem rather convincing. On the other hand, there are indications that a slowing of rate in several proteins has occurred in the primates that cannot easily be dismissed. For example, on the basis of extensive studies of globin, cytochrome c, and fibrinopeptide sequences, Goodman (1976*b*, 1981*a*) has deduced that (1) positive selection for differentiation of molecules greatly *accelerated* evolutionary rates in early vertebrates, especially at molecular sites acquiring cooperative functions; (2) stabilizing selection then preserved the functional improvements, markedly *decelerating* rates in the lineages to birds and mammals (Goodman et al., 1975); (3) *acceleration* again occurred on the step to the primates during the adaptive radiation of the mammalian orders; and (4) there was a drastic *deceleration* in evolutionary rate in "groups such as hominids" (Goodman, 1981*a*) (see Figure 8). Whether the observed variation in rates of amino acid substitution are real or are artifacts of the complicated methodology of constructing phylogenetic trees and determining the time of major

51

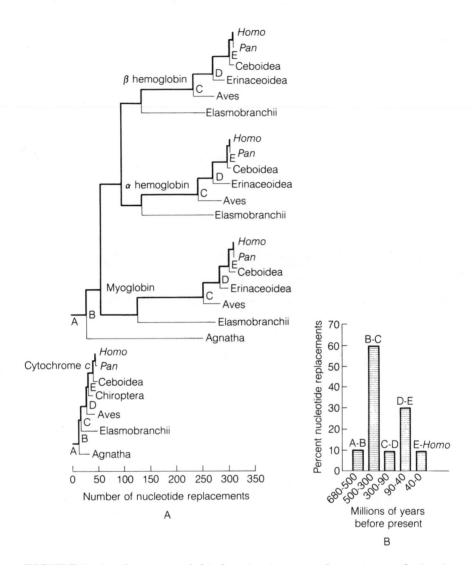

FIGURE 8. Acceleration and deceleration in rates of sequence evolution in globins and cytochrome *c*. A. Genealogical tree showing phyletic line from early vertebrates to *Homo.* Length of each branch is equivalent to average number of nucleotide substitutions, corrected for superimposed replacements, estimated to have occurred along that line of evolution. B. Bar graphs showing rates of sequence change for segments of main branches of genealogical tree shown in A. Intervals A-B, B-C, and C-D are of comparable duration; so are D-E and E-human. Note the high rate between vertebrate and bird-mammal ancestor (B–C) and again between Eutheria and Anthropoidea ancestor (D–E). (After Goodman, 1981*a*.)

branch points is open to question and is currently a topic of serious debate. In the present example, the validity of the maximum parsimony method of assigning codons and constructing genealogical trees and the augmentation procedure for correcting for hidden superimposed mutations have been questioned by several authors (Tateno and Nei, 1978, 1979; Holmquist, 1978, 1980; and Kimura, 1981*a, b*). See rebuttal by Goodman (1981*b*).

For better or worse, molecular clocks are now being widely used in studies of vertebrates, although even the staunchest advocates of the clock acknowledge that there is lineage-specific variation in rate of evolution of particular proteins, such as albumin. Among the more interesting applications is the following.

Lowenstein (1980) has been able to extract small quantities of collagen and albumin from fossils of *Australopithecus* (1.9 million years old), *Homo erectus* (500,000 years old), Neanderthals, and Egyptian mummies. These proteins react with antibodies to human collagen and human albumin, respectively, and the reactions match those observed between ordinary human collagen and its antibody and between human albumin and its antibody. The technique employed is radioimmunoassay, in which antigens or antibodies are labelled radioactively in such a way that immunological reactions to nanogram or even picogram quantities of antigen can be measured, in a scintillation counter (Prager et al., 1980).

In a recent application of this technique, Lowenstein et al. (1981) were able to determine the phylogenetic position of the Tasmanian wolf (*Thylacinus*), a marsupial that became extinct at the turn of the century. Generally, it has been placed with the Australian dasyurids (the marsupicarnivores), but a relationship to the extinct South American borhyaenids has been suggested on morphological grounds.

Lowenstein et al. (1981) extracted albumin from dried muscle adhering to a bone collected before 1893 and from two untanned skins collected early in the twentieth century. The results of testing with antisera to a variety of marsupial albumins clearly demonstrated that the Tasmanian wolf is part of a radiation leading also to *Dasyurus* and *Dasyuroides,* in the late Miocene, 6 to 10 million years ago (Figure 9). These findings agree with Simpson's (1941) and Marshall's (1977) assessments that the morphological resemblance of *Thylacinus* to the borhyaenids is due to parallel or convergent evolution.

This is the first time a phylogenetic issue has been resolved by the protein immunology of an extinct form. It would be interesting to see what the proteins of *Ramapithecus* are like!

Two calibration-free tests of the regularity of the molecular clock

53

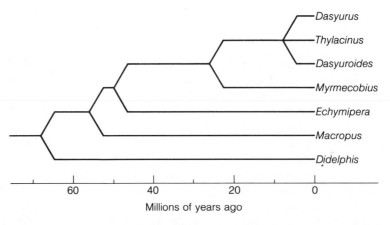

FIGURE 9. Phylogeny of the Tasmanian wolf (*Thylacinus*) and related marsupial mammals, based on radioimmunoassay of albumin. (After Lowenstein et al., 1981.)

by measuring genetic divergence between "geminate" pairs of populations of sea urchins and of marine fishes from the Atlantic and Pacific sides of the Isthmus of Panama have yielded contrary results. The rationale of the approach is that the rise of the Isthmus of Panama completely divided tropical marine organisms into Atlantic and Pacific populations. If molecular divergence occurs at a more or less constant rate, characteristic for each protein, different pairs of populations should show approximately equal degrees of molecular distance, as assessed electrophoretically over a large number of proteins. In the first of these studies, Lessios (1979) compared values of Nei's D in Atlantic and Pacific species of three genera of sea urchins, only two of which are actually relevant to the test (see Vawter et al., 1980). Species of *Eucidaris* yielded a D of 0.329, while for species of *Diadema*, $D = 0.026$. Because of this variation in D, Lessios (1979) rejected the molecular clock hypothesis, concluding that "molecules have been evolving under the influence of natural selection." But because the study was based on a small number of pairs of species, assayed for a relatively small number of enzyme loci (18 in the case of *Diadema*), rejection of the hypothesis may be premature (see critique by Vawter et al., 1980; and response by Lessios, 1981).

In a second, more ambitious study, Vawter et al. (1980) measured genetic divergence in 10 pairwise comparisons of fish populations from the Pacific and Atlantic sides of the Isthmus of Panama. Five pairs are classified as conspecific, and five are considered "geminate" species. Twenty to 40 loci were assayed electrophoretically, and Nei's D was calculated for pairs of species. By reference to the correlation between albumin immunological distance and Nei's distance demonstrated by

Sarich (1977) and modified by Carlson et al. (1978), a value of $D = 1.0$ is equal to a divergence time of 18.9 million years. The mean value of D for pairs of species was 0.2, which corresponds to 3.8 million years; the range was from 2 to 7 million years. Most geological estimates of the time of closure of the sea passage across the Isthmus of Panama fall in the range of from 2 to 5 million years ago. Hence, marine fishes are about as good as geologists in telling us when the Isthmus of Panama was formed.

LYSOZYMES

When we say that genes and proteins in two species are homologous, we normally mean that they have a degree of sequence similarity beyond that expected on the basis of chance alone. (Criteria for distinguishing homologous from analogous proteins were developed by Fitch in 1966.) But as two sequences come to differ greatly, is it still possible to tell whether the genes are homologous, that is, derived from a common ancestral gene?

Consider an enzyme that begins diverging in different lineages or, through gene duplication, in one lineage. With time, a series of changes in structure will occur, perhaps in this order:

(1) DNA sequences diverge through the accumulation of silent substitutions.

(2) Amino acid sequences diverge, as replacement mutations are fixed.

(3) Secondary and tertiary structures diverge.

(4) Interactions between substrate and enzyme diverge.

(5) Finally, mechanisms of catalysis diverge.

Over long periods, similarities in amino acid and DNA sequences may decrease to a point where the enzymes and genes can no longer be recognized as homologous by conventional criteria. But there may still be some structural correspondence, since the three-dimensional configuration of a protein changes more slowly than does its amino acid sequence.

Matthews et al. (1981) have considered the evolutionary implications of the primary, secondary, and tertiary structures of lysozyme from hen egg white and lysozyme from T4 bacteriophage. Hen lysozyme consists of 129 amino acids, with 4 disulfide linkages and no free cysteines, whereas T4 lysozyme has 164 amino acids, with no disulfide linkages and 2 free cysteines. There is no detectable amino acid sequence homology between these enzymes, but they are similar at three other levels of structure.

55

First, there is a striking resemblance in the overall three-dimensional structure of the two enzymes, and particularly in their "backbone" conformations (Figure 10). Both enzymes are bilobal, although there is much difference in pattern of folding. By allowing certain "insertions" and "deletions" of amino acids in the backbones, a three-dimensional superposition could be found in which 78 equivalent α-carbons were superimposed within a root-mean square of 4.1 Å, although this alignment did not result in any significant homology of amino acids (Rossmann and Argos, 1976; Levitt and Chothia, 1976). (Note: The three-dimensional coordinates for atoms in the polypeptides—and also for certain substrates, when in position—are known.) Figure 11 shows the 78 residues which correspond, according to the Rossmann and Argos (1976, 1978) alignment, and also the spaces between the superimposable residues, together with locations of helices and β-sheet strands. There are clearly two regions where the lysozymes superimpose over reasonably long segments: Residues 1–35

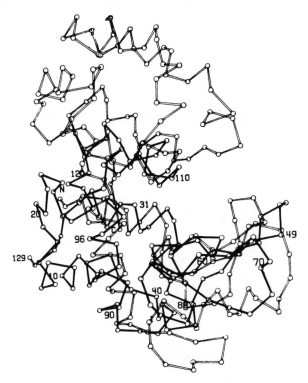

FIGURE 10. Superposition of α-carbon backbone of hen egg white lysozyme (solid lines) on that of T4 phage lysozyme (open lines). Numbers refer to amino acid sequence of hen egg white lysozyme. (After Rossmann and Argos, 1976.) See Figure 5 in Chapter 8 by Plapp.

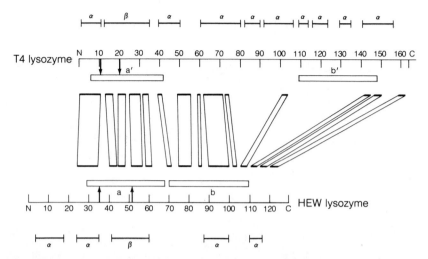

FIGURE 11. Schematic diagram showing relation between backbones of phage (T4) lysozyme and hen egg white (HEW) lysozyme. Connected solid bars indicate residues that are "equivalent" according to alignment procedure of Rossmann and Argos (1976). Arrows show locations of Glu35 and Asp52 of HEWL and Glu11 and Asp20 of T4L. Locations of α-helices and β-sheets are also indicated. Open bars a and a' are 40-residue segments which agree within 4–9 Å. Segments b and b' have the best agreement for all possible pairs of 40-residue segments, *viz,* 3–8 Å. (After Matthews et al., 1981.)

of T4L correspond (in spatial position) to 25–60 of HEWL, and residues 50–75 of T4L correspond to 75–100 of HEWL. Elsewhere the correspondence is poor, and the C terminus of T4L has no counterpart. (This is a domain of about 80 residues in T4L that appears to be involved in the recognition and binding of *E. coli* cell walls.)

Second, a striking result of the superposition work, which sought only to compare backbone configurations, is that it superimposed the active-cleft sites of the two enzymes.

Third, there are some close similarities in the details of interaction between the respective enzymes and their bound saccharide substrates (Figure 12). In particular, the pattern of hydrogen bonding between the substrate and enzyme is strikingly similar. Note the hydrogen bonds between the acetamido group of the substrate molecule and the main chain carbonyl ($C=O$) and NH groups of Alanine 107 and Asparagine (*Asn*) 59, respectively, in HEWL. In phage lysozyme, there are similar bonds, but to Phenylalanine 104 and Leucine 32. Other hydrogen bonds between substrate and protein are dissimilar.

57

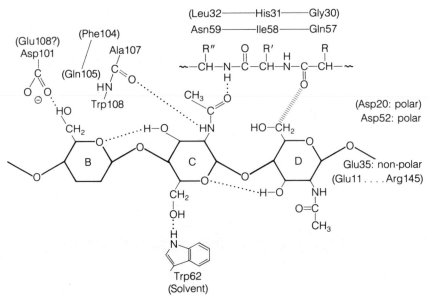

FIGURE 12. Schematic drawing comparing saccharide-protein interactions in phage lysozyme (names in parentheses) and hen egg white lysozyme. Dotted lines illustrate apparent close contact that occurs when saccharide D is in normal chair conformation. (After Matthews et al., 1981.)

Are the structural similarities between hen egg-white lysozyme and phage lysozyme convergent, or do they stem from descent from a common ancestral molecule? Matthews et al. (1981) conclude that these lysozymes have in fact evolved from a common precursor. This is the first example of two proteins whose amino acid sequences are completely dissimilar but for which a reasonable case can be made for homology. Even if one is not convinced by the argument for homology, one must agree that the approach is clever and promising.

With this example, we have come full circle, working from the morphological phenotype, without any useful knowledge of the gene, through the primary structure (amino acid sequences) of proteins to the nucleotide structure of the gene itself, and finally back to purely phenotypic structural considerations. The coordinate systems of Matthews et al. (1981) are vaguely reminiscent of those long ago used by D'Arcy Thompson in studying morphological resemblance and transformation.

FINAL COMMENTS

In 1945, G. G. Simpson concluded that, "The stream of heredity makes phylogeny; in a sense, it is phylogeny. Complete genetic analysis

would provide the most priceless data for the mapping of this stream."
We now know that complete genetic analysis is the specification of
nucleotide sequences of nuclear and extra-nuclear DNA (or RNA).
This has already been achieved for several viruses, and also for mitochondria, genes determining globins, genes determining interferons,
other genes in mammals, and for many genes in bacteria. Phylogenetic
trees based on nucleotide sequences will soon replace those derived
from amino acid sequences, which are more difficult to obtain and
carry less phylogenetic information. Considering the major contributions already made by nucleotide sequencing to the study of gene
evolution and phylogeny, it takes little imagination to predict that
the continuing study of the structure of the gene at the most fundamental level will soon tell us more about the phylogenetic relationships of organisms than we have managed to learn in all the 173 years
since Lamarck.

CHAPTER 4

THE GENETIC STRUCTURE
OF SPECIES

Francisco J. Ayala

THE PROBLEM

In sexually reproducing organisms, species may be defined as groups of interbreeding natural populations that are reproductively isolated from other such groups. A species is a natural unit or system, defined by the possibility of interbreeding between its members; the ability to interbreed is of considerable evolutionary import, because it establishes species as discrete evolutionary units. A gene mutation or some other genetic change originating in a single individual may spread, over the generations, to all members of the species but ordinarily not to individuals of other species. Owing to reproductive isolation, different species have separately evolving gene pools.

Hereditary changes underlie the evolutionary process. The evolutionary potential of a population is a function of the amount of genetic variation present in a population (and also, of course, of the rate at which new genetic variation can be generated by the mutation process, but this rate will largely be reflected in the amount of genetic variation existing at a given time). This is simply because the greater the number of variable gene loci and the more alleles there are at each variable locus, the greater the possibility for change in the frequency of some alleles at the expense of others. How much genetic variation exists in populations is, then, of fundamental importance for the understanding of the evolutionary process.

The study of genetic variation in natural populations is characterized by a gradual discovery of ever increasing amounts of genetic variation. In the early decades of this century, geneticists thought that an individual is homozygous at most gene loci and that individuals of the same species are genetically almost identical. Recent discoveries suggest that, at least in outcrossing organisms, the DNA sequences inherited one from each parent are likely to be different for every gene locus in every individual. That is, every individual may be heterozygous at every gene locus. But this is, at present, only a conjecture because of the limited evidence available, and we are, in any case, a long way from knowing with any degree of precision the amount of variation in DNA sequence in any organism.

In the following pages, I will summarize the knowledge acquired over the last few decades concerning the genetic structure of species. The question of how much genetic variation exists in populations will be considered in some detail. How the variation existing in a species is distributed among the local populations that make up the species will be considered only briefly in the last part of this chapter.

It should be made clear from the beginning that I am here concerned exclusively with descriptive questions, namely with the questions of how much variation there is in a species and how it is organized within the species. Questions about causes will not be discussed. The investigation of the processes by which genetic variation arises and is preserved or increased is of great evolutionary interest, but it is not part of the present subject.

FROM THE HAZY TO THE DELUSIVE

Genetic variation is an attribute that cannot be exhaustively measured. It is not possible, even if we wanted it, to examine every gene in every individual of a given species, so as to obtain a complete enumeration of the genetic variation in the species. The solution in such a situation is, of course, to measure a sample from the group to be evaluated. Two conditions need to be met for a valid extension of the results obtained in the study of a sample to the whole set. First, the sample must be representative or *unbiased*; second, the sample must be *accurately* measured. In the case at hand, the requirement that the sample be unbiased applies to two levels: (1) the individual organisms sampled must be, on the average, neither more nor less genetically variable than the population as a whole; (2) the genes sampled must be neither more nor less polymorphic, on the average, than the whole genome. And the condition of accuracy requires that genes that are different be identified as such; i.e., it requires that every allelic variant be recognizable.

Neither one of the two necessary conditions for valid sampling have been met in the study of genetic variation. There is no serious difficulty in sampling *individuals* that are, on the average, as genetically variable as the population as a whole. An important consideration is that the individuals sampled not be particularly either inbred or interrelated; but this is not difficult to satisfy. The difficulty lies in choosing the *genes* to be sampled. With the methods of Mendelian genetics, the existence of a gene is ascertained by examining the progenies of crosses between individuals showing different forms of a given character; from the proportion of individuals in the various classes, we infer whether one or more genes are involved. By such methods, therefore, the only genes known to exist are those that are variable. There is no way of obtaining an unbiased sample of the genome, because invariant genes cannot be included in the sample of genes to be examined.

A way out of this problem became possible with the advent of molecular genetics. It is now known that the genetic information encoded in the coding sequence of the DNA of a structural gene is translated into a sequence of amino acids making up a polypeptide. One can select for study a series of proteins without previously knowing whether or not they are variable in a population—a series of proteins that, with respect to variation, are an unbiased sample of all the structural genes in the organism. If a protein is found to be invariant among individuals, it is inferred that the gene coding for the protein is also invariant; if the protein is variable, the gene is inferred also to be variable and one can measure how variable it is, i.e., how many variant forms of the protein exist, and in what frequencies.

Gel electrophoresis is a fairly simple technique that makes possible the study of protein variation with only a moderate investment of time and money. Since the 1960s, genetic variation has been studied in a large variety of organisms by gel electrophoresis. It was clear from the beginning of these studies that not all allelic variants are detected by electrophoresis, and hence that the second necessary condition for measuring genetic variation is not satisfied. But genes are selected for electrophoretic studies because simple techniques exist for assaying the proteins they encode. Because the genes are chosen without regard to how variable they are, many investigators thought that electrophoresis would provide estimates of variation in structural genes that would be accurate to a first approximation. Moreover, in their more optimistic moments, some evolutionists expected that some way would be found to evaluate to what extent electrophoresis underestimates genetic variation; simple relationships might be discovered that would permit transforming electrophoretic estimates into fairly accurate measures of genetic variation.

However, these expectations have not been realized. At present, it appears doubtful that the genes studied by electrophoresis are an unbiased sample of the structural genes, let alone the genome as a whole. It is questionable whether formulae can be found (or even whether they are worth finding) to transform electrophoretic measures into "true" estimates of genetic variation even for the genes assayed by electrophoresis. Because of the pervasive enthusiasm originally generated by the electrophoretic studies, the decade from the mid-1960s to the mid-1970s may be called the "romantic period" in the study of genetic variation. One might also call it the "delusive period," because the anticipated expectations remain largely unfulfilled.

The past few years have witnessed a new important development: techniques for the isolation ("cloning") of genes and other DNA segments and for ascertaining their nucleotide sequence. The condition of accurate measurement is fully satisfied by these techniques, because every nucleotide difference (i.e., every allele) can be detected. There is hope that the condition of unbiased sampling may also be satisfied, because all sorts of genes, whether translated or only transcribed, and indeed any kind of DNA sequence can be subject to study. We are, thus, in a "period of great expectations." Only the future will tell whether these expectations are fulfilled.

EARLY EVIDENCE OF EXTENSIVE VARIATION

The evidence for genetic variation can be traced to Mendel's experiments: the discovery of the laws of heredity was made possible by the expression of segregating alleles. Gradually, it became apparent that genetic variation is pervasive. Until the mid-1960s the evidence for this pervasiveness came primarily from three kinds of study: morphological variation, artificial selection, and inbreeding.

Morphological variation among individuals is a conspicuous phenomenon whenever organisms of the same species are carefully examined. Human populations, for example, exhibit variation in facial features, skin pigmentation, hair color and shape, body configuration, height and weight, etc. A difficulty is that it is not apparent how much of this variation is due to genetic variation and how much to environmental effects. The genetic basis of morphological variants has been ascertained in a number of favorable cases, such as shell banding in snails and flower color in plants. But it is not generally known how many genes may affect the morphology of an organism or what proportion of them are polymorphic.

A convincing source of evidence indicating that genetic variation is pervasive comes from artificial selection experiments. Artificial selection has been successful for innumerable commercially desirable traits in many domesticated species, including cattle, swine, sheep, poultry, corn, rice, and wheat, as well as in many experimental organisms, such as mice and *Drosophila* flies. In *D. melanogaster* artificial selection has been successful for more than 50 different traits. Artificial selection can only succeed if the original population has genetic variation with respect to the trait being selected. The fact that artificial selection succeeds virtually every time that it is tried suggests that genetic variation is common in natural populations for the kinds of traits, such as size, fertility, etc., that are selected. But it is impossible to say what proportion of the genome may be involved in these traits.

Throughout the 1950s and 1960s ingenious and laborious experiments were carried out in *Drosophila* to reveal by inbreeding the presence of genetic variants that are deleterious in homozygous condition, but not in heterozygotes (Dobzhansky et al., 1977). The techniques essentially consist of obtaining numerous flies all homozygous for a given chromosome sampled from a natural population. Recessive allelic variants present in the wild chromosome will be expressed in the homozygous flies. The joint effects on fitness of all the genes in the chromosome can be measured as the average performance of the set of homozygous flies relative to "normal" flies, i.e., flies heterozygous for random combinations of wild chromosomes. The parameter measured in most experiments has been viability from egg to adult.

Typical results are summarized in Table 1. Between one-quarter and one-third of all the chromosomes carry one or more genes that cause lethality in homozygous condition. In addition, about 50 percent of the chromosomes significantly reduce the viability of homozygotes.

TABLE 1. Frequencies of wild chromosomes of *Drosophila pseudoobscura* from the Sierra Nevada, California, that in homozygous condition have various effects on viability. (After Dobzhansky and Spassky, 1953, 1963)

Effect on viability	FREQUENCY (%) OF CHROMOSOME		
	Second	Third	Fourth
Lethal or semilethal	33.0	25.0	25.9
Subvitality (significantly lower viability than wild flies)	62.6	58.7	51.8
Normal (viability not significantly different from wild flies)	4.3	16.3	22.3
Supervitality (significantly higher viability than wild flies)	<0.1	<0.1	<0.1

TABLE 2. Variation in viability obtained through recombination between chromosomes having "quasinormal" viability in homozygous condition. (After Dobzhansky et al., 1959)

Species	Lethal chromosomes	VARIANCE OF VIABILITY		
		Among natural chromosomes (a)	Among recombinant chromosomes (b)	Recovered variance (b/a)
Drosophila pseudoobscura	4.1%	140	60	42.9%
Drosophila prosaltans	5.7%	200	50	25.0%

It might be assumed that the chromosomes that give homozygotes with normal viability are normal chromosomes, all with the same genetic constitution and thus lacking genetic variation affecting viability. However, this suggestion has been refuted by two kinds of experiments. First, recombinant chromosomes are obtained in the progeny of flies heterozygous for such "normal" chromosomes. When these recombinant chromosomes are again made homozygous, it is found that they are heterogeneous, some being lethal, others semilethal, and still others normal (Table 2). The variance among the recombinant chromosomes is still substantial, showing that "normal" chromosomes that do not reduce viability are far from genetically uniform.

The second kind of experiments measure the fitness of homozygous flies in population cages over a number of generations. The chromosomes that yield normal viability in the type of experiment just discussed turn out to be extremely heterogeneous with respect to various fitness components. A relevant result of this second type of experiment is that when *total fitness* is measured, no chromosome is normal in homozygous condition (Table 3). Even the chromosomes that appear normal when only viability is measured, become genetically lethal or semilethal when reproduction is taken into account. The conclusion is that variability with respect to fitness is present in every chromosome. But, how many genes are involved? Tracey and Ayala (1974) have concluded that the results are consistent with the occurrence of more than one thousand polymorphic loci in natural populations of *Drosophila*. But the actual number cannot be established, not even approximately, by these experiments.

TABLE 3. Fitness of *Drosophila* homozygotes for quasinormal chromosomes.

Species	Chromosome	Sample size	Average fitness	Reference
D. pseudoobscura	II	16	0.37	Sved and Ayala, 1970
D. melanogaster	X	34	0.40	Wilton and Sved, 1979
	II	24	0.15	Sved, 1971
	II	23	0.19	Tracey and Ayala, 1974
	III	14	0.32	Tracey and Ayala, 1974
	III	14	0.10	Sved, 1975
	II & III	24	0.08	Seager, 1979
D. willistoni	II	15	0.34	Mourão et al., 1972

THE ELECTROPHORESIS REVOLUTION

Gel electrophoresis followed by selective staining provides a simple method for assaying variation in enzymes and other soluble proteins. Most of the proteins assayed are encoded by single-gene loci. The gel patterns can, then, be interpreted as single-locus genotypes. Genotypic and allelic frequencies, as well as other relevant genetic information, can be readily obtained.

Electrophoresis makes it possible to study gene loci independently of whether they are variable or not (Figure 1). Thus, the way might seem open for obtaining measures of genetic variation, even though these measures are *minimum* estimates because not all allelic differences are detected by electrophoresis.

The application of electrophoretic techniques to the study of genetic variation generated enormous enthusiasm among evolutionists for one additional reason: it provides a method for obtaining genetic information from organisms not suitable for breeding experiments. Organ-

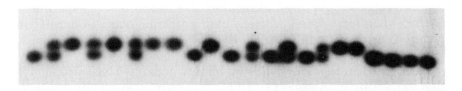

FIGURE 1. An electrophoretic gel stained for the enzyme phosphoglucomutase. The gel contains tissue samples from each of 22 females of *Drosophila pseudoobscura*. A fly exhibiting only one colored spot is concluded to be homozygous; the first and third flies from the left are homozygous for different alleles. A fly showing two spots, like the second one from left, is heterozygous.

isms with long generations or those that cannot be bred in the laboratory, because they live in exotic environments (such as the deep sea) or for other reasons, can now be assayed for certain genetic parameters.

Before the electrophoretic revolution, genetic data existed for only a few dozen multicellular organisms. Now, hundreds of different species have been studied by electrophoresis. The number of loci sampled in many species is sufficiently large, 15 or more, so that average estimates of genetic variation can be advanced with some degree of confidence. A partial summary is given in Table 4; reviews can be found in Gottlieb (1981), Koehn and Eanes (1978), Nevo (1978), Selander (1976), and Powell (1975).

TABLE 4. Genetic variation in natural populations of some major groups of animals and plants. (From Ayala and Kiger, 1980)

Organisms	Number of species	Average number of loci per species	Average polymorphism*	Average heterozygosity
INVERTEBRATES				
Drosophila	28	24	0.529	0.150
Wasps	6	15	0.243	0.062
Other insects	4	18	0.531	0.151
Marine invertebrates	14	23	0.439	0.124
Land snails	5	18	0.437	0.150
VERTEBRATES				
Fishes	14	21	0.306	0.078
Amphibians	11	22	0.336	0.082
Reptiles	9	21	0.231	0.047
Birds	4	19	0.145	0.042
Mammals	30	28	0.206	0.051
PLANTS				
Self-pollinating	12	15	0.231	0.033
Outcrossing	5	17	0.344	0.078
OVERALL AVERAGES				
Invertebrates	57	22	0.469	0.134
Vertebrates	68	24	0.247	0.060
Plants	17	16	0.264	0.046

* The criterion of polymorphism is not the same for all species.

67

Electrophoretic data give the frequency of electromorphs (proteins that differ in electrophoretic mobility). Proteins encoded by different alleles may yield indistinguishable electromorphs, but, as a first approximation, it is assumed that each electromorph corresponds to only one allele. A variety of statistics can be used to summarize the amount of genetic variation in a population. The most extensively used measures are the *polymorphism* (*P*) and the *heterozygosity* (*H*). *P* is simply the proportion of loci found to be polymorphic in the sample. Usually, a locus is considered polymorphic when the frequency of the most common allele (electromorph) is no greater than a certain value, such as 0.99 or 0.95. In outcrossing organisms, *H* estimates the average frequency of heterozygous loci per individual or, what is equivalent, the average frequency of heterozygous individuals per locus. In naturally inbred organisms, *H* is a good measure of genetic variation in a population, if it is calculated from the allelic frequencies as the "expected" frequency of heterozygous individuals on the assumption of Hardy-Weinberg equilibrium. *H* is a better measure of genetic variation than *P* for most purposes, because it is more precise (Dobzhansky et al., 1977).

Electrophoretic studies have confirmed that natural populations of most organisms possess large stores of genetic variation. Even though not all variants are detected, Table 4 shows that the average heterozygosity is between 4 and 8 percent for a given vertebrate group, and between 6 and 15 percent for a given invertebrate group. Even self-fertilizing plants have considerable genetic variation. The average proportion of polymorphic loci in a population lies between 20 and 50 percent for most organisms.

There is, however, considerable heterogeneity in the amount of genetic variation among organisms within a given group—a fact concealed in Table 4, where only averages are given. Among fish, for example, $\bar{H} = 0.005$ in *Trematomus borchgrevinki*, but $\bar{H} = 0.107$ in *Dascyllus reticulatus* (Somero and Soulé, 1974). Within a single genus of krill, $\bar{H} = 0.057$ in *Euphausia superba,* but $\bar{H} = 0.213$ in *E. distinguenda* (Ayala and Valentine, 1979). Moreover, differences in the amount of genetic variation also exist among local populations of the same species (see below).

THE PROBLEM OF ACCURACY

How accurate are electrophoretic estimates? That is, what proportion of the total variation is detected by electrophoretic techniques? Electrophoresis cannot, of course, detect nucleotide substitutions that do not change the encoded amino acids. The question is what proportion of amino acid substitutions are detected. Some biologists have argued that electrophoresis detects only substitutions that change the net electric charge of the encoded proteins and have calculated that about 67 percent of all amino acid substitutions are electrophoretically cryptic (see Marshall and Brown, 1975). It is now known, however, that electrically neutral charges can be detected, at least in some cases (e.g., Ramshaw et al., 1979).

The question raised could ultimately be resolved by obtaining the amino acid sequence of a sufficiently large number of electromorphs with identical electrophoretic mobility. This is clearly not feasible at present because of the enormous cost and time required. A variety of other, less satisfactory, methods have manifested the existence of electrophoretically cryptic variation. The methods used include sequential electrophoresis, heat denaturation, urea denaturation, monoclonal antibodies, and peptide mapping.

Electrophoretic studies of genetic variation usually employ a single set of experimental conditions in the assay of a given enzyme. The method of sequential electrophoresis consists of applying a variety of conditions to a given enzyme. The conditions most often varied are gel concentration and buffer pH; typically from six to ten different sets of conditions are used. Electromorphs that have identical mobility under a set of conditions may be distinguishable when the conditions are changed.

The species most extensively examined by sequential electrophoresis is *Drosophila pseudoobscura*. Table 5 summarizes some of the results obtained (some data have not been included because the statistics used in the table could not be calculated). H is the frequency of heterozygous individuals; n_e is the effective number of alleles. A prime mark (H', n_e') is used to distinguish the values obtained by sequential electrophoresis from those based on standard conditions. The increase in variation detected ranges from zero, at a number of loci, to 76 percent at the *Est-5* locus ($n_e'/n_e = 1.76$). In general, the more frequently heterozygous the locus, the greater the increase in variation appears to be. However, this rule has exceptions. For example, although *Adh* in *D. melanogaster* is very frequently heterozygous ($H \approx 0.50$), no additional variation has been detected by sequential electrophoresis (Kreitman, 1980). For the 13 loci shown in Table 5, there is an average increase of 12 percent ($n_e'/n_e = 1.12$) in the amount of variation; with H increasing by 0.04. The average heterozygosity for the 13 loci is greater than it has been observed in studies of *D. pseudoobscura* that include a larger number of loci ($\overline{H} \approx 0.10$). Thus, if the increase in variation detected by sequential electrophoresis is proportional to H, then the average increase for random samples of electrophoretic loci would be even less than indicated in Table 5. Be

TABLE 5. Increase in genetic variation observed by sequential electrophoresis in U.S. populations of *Drosophila pseudoobscura*.

Locus	STANDARD CONDITIONS		ALL CONDITIONS		INCREASE IN VARIATION		Reference
	H	n_e	H'	n'_e	$H' - H$	n'_e/n_e	
Est-5	0.645	2.8	0.798	4.9	0.153	1.76	Coyne et al., 1978
Pt-8	0.51	2.04	0.55	2.22	0.04	1.09	Singh, 1979
Ao*	0.499	2.0	0.584	2.4	0.085	1.20	Coyne and Felton, 1977
Xdh	0.436	1.8	0.628	2.7	0.192	1.50	Coyne and Felton, 1977
Odh	0.082	1.09	0.098	1.11	0.016	1.02	Coyne and Felton, 1977
Hex-1	0.077	1.08	0.077	1.08	0.00	1.00	Beckenbach and Prakash, 1977
Pt-7	0.06	1.06	0.09	1.10	0.03	1.03	Singh, 1979
Pt-6	0.04	1.04	0.04	1.04	0.00	1.00	Singh, 1979
Mdh	0.00	1.00	0.00	1.00	0.00	1.00	Coyne and Felton, 1977
Hex-2	0.00	1.00	0.00	1.00	0.00	1.00	Beckenbach and Prakash, 1977
Hex-6	0.00	1.00	0.00	1.00	0.00	1.00	Beckenbach and Prakash, 1977
Hex-7	0.00	1.00	0.00	1.00	0.00	1.00	Beckenbach and Prakash, 1977
α Gpdh	0.00	1.00	0.00	1.00	0.00	1.00	Coyne et al., 1979
Average	0.181	1.38	0.221	1.65	0.040	1.12	

* Labeled *Adh-6* in the original paper.

TABLE 6. Increase in genetic variation observed by heat denaturation in *Drosophila melanogaster*.

Locus	ELECTROPHORESIS		ELECTROPHORESIS PLUS HEAT		INCREASE IN VARIATION		Reference
	H	n_e	H'	n_e'	$H' - H$	n_e'/n_e	
Est-6	0.48	1.93	0.67	2.98	0.19	1.54	Cochrane and Richmond, 1979
Adh	0.48	1.94	0.50	2.00	0.02	1.03	Sampsell, 1977
Pgm	0.36	1.57	0.45	1.82	0.09	1.16	Trippa et al., 1976
α Gpdh	0.32	1.47	0.32	1.47	0.00	1.00	Sampsell, 1977
Average	0.410	1.73	0.485	2.06	0.075	1.18	

that as it may, the increase detected by sequential electrophoresis, in either heterozygosity or in the effective number of alleles, is relatively small.

Electromorphs with identical electrophoretic mobility may differ in thermostability. If the differences are shown by genetic tests to be associated with the locus itself, they may reflect different amino acid sequences in the encoded polypeptides. Satisfactory data exist for four enzymes in a single species, *D. melanogaster* (Table 6). All four loci are highly polymorphic, but the additional variation detected ranges from none ($\alpha Gpdh$) to 54 percent (*Est-6*). On the average, $\overline{H}$ increases from 0.410 to 0.485 and the effective number of alleles is 18 percent greater. The increase in variation is somewhat larger than with sequential electrophoresis, but this should be taken under advisement, given the paucity of the data.

Loukas et al. (1981) have assayed by urea denaturation eight loci in *D. subobscura*. Differences in sensitivity to urea treatment appear to be associated with each locus under examination rather than with the genetic background. The results are summarized in Table 7. There is some association between the amount of electrophoretic variation and the increase in variation detected by urea, although *Pept-1* is a notable exception. The average increase in heterozygosity (0.077) is about the same as the average increase detected by thermostability in *D. melanogaster,* but the increase in effective number of alleles is greater (25 percent vs. 18 percent). Satoh and Mohrenweiser (1979) have suggested that the variation uncovered by urea denaturation may be the same as that uncovered by thermostability tests (see also Loukas et al., 1981).

TABLE 7. Increase in genetic variation observed by urea denaturation in two Greek populations of *Drosophila subobscura*. (After Loukas et al., 1981)

Locus	ELECTRO-PHORESIS		ELECTRO-PHORESIS PLUS UREA		INCREASE IN VARIATION	
	H	n_e	H'	n'_e	$H' - H$	n'_e/n_e
Xdh	0.61	2.56	0.77	4.44	0.16	1.73
Est-5	0.60	2.52	0.73	3.74	0.13	1.48
Ao	0.59	2.41	0.69	3.19	0.10	1.32
Pept-1	0.48	1.95	0.48	1.95	0.00	1.00
Lap	0.47	1.90	0.62	2.67	0.15	1.41
Odh	0.12	1.14	0.12	1.14	0.00	1.00
Me	0.10	1.11	0.11	1.13	0.01	1.02
Acph	0.06	1.06	0.13	1.15	0.07	1.08
Average	0.379	1.83	0.456	2.42	0.077	1.25

TABLE 8. Increase in genetic variation detected by three different methods at the *Adh* locus of *Drosophila melanogaster*.

Method	$H' - H$	n'_e/n_e	Reference
Sequential electrophoresis	0.00	1.00	Kreitman, 1980
Heat denaturation	0.02	1.03	Sampsell, 1977
Peptide mapping	0.10	1.20	Fletcher, 1979

Slaughter et al. (1981) have used a panel of six monoclonal antibodies to detect genetic variation in human alkaline phosphatase. The variants detectable by electrophoresis (three common and about 20 rare alleles) are also discriminated by this method. In addition, at least one more allele has been detected: one electromorph class turns out to consist of two alleles distinguishable by their antigenic properties. Immunological assay by means of monoclonal antibodies appears to be, thus, another method to uncover genetic variation. But it is not possible to draw any conclusions as to its effectiveness, because the only enzyme studied has not been surveyed by other methods for detecting electrophoretically cryptic variation.

Peptide mapping or "fingerprinting" uses electrophoresis and chromatography to separate, in two dimensions, soluble peptides obtained by enzymatic digestion of a polypeptide. If the peptides obtained by digestion are sufficiently small, it seems likely that virtually every amino acid substitution would be detectable. Purifying and fingerprinting a protein is, however, a laborious process. Fletcher (1979) has examined 11 separate allelic products of *Adh* in *D. melanogaster*. Five code for the slow (S) electromorph and yield identical peptide maps. The other six code for the fast (F) electromorph: one of these six is different from the others. Thus, the frequency of electrophoretically cryptic variants in this sample is $1/11 = 0.091$. In the populations sampled, $H = 0.42$ and $n_e = 1.72$ for electrophoretically detectable variation. After peptide analysis, $H' = 0.52$ and $n'_e = 2.07$, or an increase of 0.10 in heterozygosity and of 20 percent in the effective number of alleles.

The *Adh* locus of *D. melanogaster* has been assayed by sequential electrophoresis and heat denaturation, as well as by peptide mapping. The results are compared in Table 8. Using sequential electrophoresis, Kreitman (1980) failed to detect any electrophoretically cryptic variants in a sample of 96 allelic products (although he could differentiate some of the variants first identified by thermostability). Sampsell (1977) examined by heat denaturation 4,436 allelic products and ob-

73

tained a small increase in variation. Although the allelic sample studied is much too small, peptide analysis appears to have considerably greater power for detecting electrophoretically cryptic variation than the two other techniques. Yet, it corroborates the results obtained by these techniques in the sense that it suggests that conventional electrophoresis may actually detect most of the protein variation present in natural populations. An increase of 20 percent in the effective number of alleles is not trivial, but it is not likely to have drastic consequences for most evolutionary considerations.

VARIATION IN REGULATORY GENES

Accuracy—ability to discriminate among all allelic products that are different—is one of the conditions that data must meet in order to obtain valid estimates of genetic variation. The other condition is lack of bias, i.e., the genes studied must be a random representation of all the genes in the organism. The genes surveyed by electrophoresis are structural genes coding for soluble proteins. Whether all such genes are randomly represented in electrophoretic surveys is a question that cannot be answered at present. It is not known, either, whether structural genes coding for nonsoluble proteins are either more or less variable than genes coding for soluble proteins. The large majority of the DNA of eukaryotic organisms, however, does not code for proteins. Much of this additional DNA may primarily, or exclusively, have a structural function, but a fraction is involved in gene regulation and may be of considerable evolutionary import.

Regulatory genes, *sensu lato,* are those that regulate or modify the activity of other genes (Ayala and Kiger, 1980). Thus defined, regulatory genes may include genes that code for proteins. Nevertheless, it remains important to ascertain the extent of genetic variability in the regulation of structural genes.

Several studies have shown that the level of ADH activity in *D. melanogaster* can be affected by genes other than *Adh,* the locus which codes for the protein (Ward and Hebert, 1972; Pipkin and Hewitt, 1972; Ward, 1975; McDonald et al., 1977). The *Adh* locus is in the second chromosome of *D. melanogaster.* McDonald and Ayala (1978) have shown that the third chromosome contains genes that regulate the activity of the *Adh* locus and that there is considerable genetic variation in natural populations with respect to these third-chromosome regulatory genes. The tests consist in combining, in homozygous condition, a given second chromosome with each of several third chromosomes. The results are shown in Table 9. The effects of the third chromosomes are considerable. For example, when the second chromosome labelled 6S is combined with each of five different third chromosomes, ADH activity ranges from 7.9 to 24.7, a threefold increase.

TABLE 9. Variation in ADH activity due to regulatory genes located on the third chromosome of *Drosophila melanogaster*. The *Adh* locus is on the second chromosome. The data given measure levels of enzyme activity in flies homozygous for both the second and the third chromosome. (After McDonald and Ayala, 1978)

Second chromosome	THIRD CHROMOSOME								Activity effects of third chromosomes*
	1F	**2F**	**3F**	**1S**	**2S**	**3S**	**5S**	**6S**	
1F	18.4	25.7	—	—	—	19.9	—	19.1	2F ≫ 3S = 6S = 1F
2F	—	25.5	44.4	22.5	—	24.5	21.5	—	3F ≫ 2F = 3S > 1S = 5S
3F	—	—	23.5	15.4	—	—	16.4	—	3F ≫ 5S = 1S
1S	—	—	16.0	8.9	14.8	—	—	10.4	3F ≫ 2S ≫ 6S > 1S
2S	—	10.9	28.9	—	8.9	—	—	—	3F ≫ 2F = 3S
3S	11.0	7.0	—	—	—	5.8	—	5.4	1F ≫ 2F = 3S = 6S
6S	—	17.8	24.7	7.9	—	—	22.4	10.6	3F ≫ 5S ≫ 2F ≫ 6S ≫ 1S

* The meaning of the symbols is: =, not significantly different; >, greater than, with P < 0.05; ≫, greater than, with P < 0.01.

C. Laurie-Ahlberg and A. Wilton (personal communication) have used techniques similar to those of McDonald and Ayala (1978) in a survey of 23 enzymes in *D. melanogaster*. The activity of 19 of the enzymes (83 percent) is significantly affected by genetic variation in chromosomes other than that in which the gene coding for the enzyme is located; the effects on three other enzymes are marginally significant.

Variation in regulatory or modifier genes affecting the activity of other genes is, therefore, pervasive. Unfortunately we do not know how many gene loci may have modifier effects on any given enzyme locus. At present there seems to be no way in which the variation in gene regulation can be quantified using statistics such as H or n_e. With respect to variation in gene regulation, we are at the same stage where we were with respect to structural genes before the use of electrophoretic techniques. We can state that the variation is extensive, but we cannot tell how many regulatory genes are polymorphic or how polymorphic they are.

VARIATION IN DNA SEQUENCE

It has been known for more than a decade that only a small fraction, perhaps less than 10 percent, of the nuclear DNA of eukaryotes is

75

translated into protein. The recently developed techniques of DNA cloning and sequencing have shown that genes are separated from each other by long DNA sequences that do not become transcribed into RNA. The genes themselves have a complex organization. At both ends they have relatively short sequences that are present in the mature mRNA transcript, but do not code for amino acids. Most genes contain, in addition, intervening sequences (introns), which separate from each other the segments that code for the amino acids (exons). The introns are transcribed in the nucleus, together with the rest of the gene, but they are spliced out before the mRNA migrates to the cytoplasm (Figure 2).

The question of how much genetic variation exists in the DNA of an organism can, thus, be formulated in various ways. One may ask about the whole genome or about particular components such as, for example, the coding segments. A number of genes have been sequenced in two or more related species, and it has become apparent that different segments evolve at different rates. This suggests that different kinds of segments may have different levels of polymorphism, a hypothesis recently corroborated by direct evidence.

Slightom et al. (1980) have sequenced two alleles of the $^A\gamma$ gene, which codes for one of the polypeptides of fetal hemoglobin. The two alleles are from a single individual, one allele from the paternal and the other from the maternal chromosome. The results are summarized

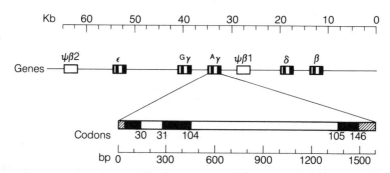

FIGURE 2. Organization of the beta family of human globin genes. These genes are located in chromosome 11. Each functional gene consists of three exons (black) separated by two introns (white). The "pseudogenes" $\psi\beta1$ and $\psi\beta2$ are represented by white rectangles. In the $^A\gamma$ gene, one intron separates the triplets coding for amino acids 30 and 31; the other is between triplets for amino acids 104 and 105. The mature mRNA transcript of each gene includes the three exons plus the untranslated sequences at each end, shown by hatching. The DNA segment represented on top is 62 kilobases (Kb, thousand base pairs) in length. Each gene consists of about 1,600 base pairs (bp); 438 of these code for the 146 amino acids of each polypeptide. (After Slightom et al., 1980.)

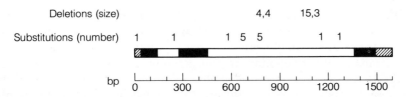

FIGURE 3. Diagram showing the local distribution of nucleotide differences between two allelic $^A\gamma$ genes. All differences are in untranslated DNA. The number of nucleotide substitutions per hundred-base interval is given; the four deletions are characterized according to size in base pairs. See Figure 2.

in Figure 3. The three exons have identical sequences, but nucleotide substitutions occur in the 5′ flanking sequence and in both introns. Most of the substitutions occur in the 5′ region of the larger intron; the two alleles also differ in this region by two four-nucleotide gaps. The length of the DNA sequenced is 1,468 nucleotide pairs; the total number of substitutions is 24. Depending on whether or not the two gaps are counted as differences, the percent of nucleotide differences is either 2.2 or 1.6.

The constant region of the heavy chain of mouse immunoglobulin consists of eight proteins. One of these, γ2a, is known to differ extensively from one inbred mouse strain to another. The gene, IgG2a, coding for this protein has been sequenced in two strains (Schreier et al., 1981). Of the 1,108 bases sequenced, 111 (10 percent) are different. Only 18 (16.2 percent) of these nucleotide substitutions are silent; the others result in substitutions of 15 percent of the amino acids.

There are reasons to presume that the variation observed in the mouse IgG2a gene may not be typical of structural loci. Immunoglobulin genes are very polymorphic; the two alleles sequenced come from two inbred strains, rather than from outbred individuals; the two proteins were known to be very different before the DNA was sequenced. Indeed the frequency of amino acid differences between the two allele products is one order of magnitude greater than the average observed in other kinds of protein.

In any case, the globin gene results obtained by Slightom et al. (1980) suggest that, at least if introns are taken into account, every diploid individual may be heterozygous at virtually every gene locus. When the DNA base-sequence is considered, questions about heterozygosity will have to be answered not in terms of gene loci (because 100 percent of the loci are likely to be heterozygous) but in terms of nucleotides. And there is evidence indicating that the values reported above—2 percent and 10 percent nucleotide differences for the globin and the immunoglobulin gene, respectively—might not be far off the mark.

The genome of eukaryotes consists of single-copy DNA, which typically may be around 70 percent of the total, and of repetitive DNA. The latter is made up of sequences each represented by several copies, sometimes many thousands, in the genome. Britten et al. (1978) and Grula et al. (1982) have used techniques for DNA denaturation, followed by competitive reassociation ("hybridization") of the dissociated DNA strands, in order to estimate the amount of nucleotide variation in single-copy DNA. The estimated frequencies of nucleotide substitutions in the four species of sea urchins examined are *Strongylocentrotus purpuratus,* 4 percent; *S. franciscanus,* 3.2 percent; *S. intermedius,* 3 percent; and *S. droebachiensis,* 2 percent. The single-copy DNA consists of two fractions, one less polymorphic than the other. The less polymorphic fraction makes up the larger part of the DNA. In *S. purpuratus* the "heterozygosity" values are 3 percent and 9 percent for the less polymorphic and the more polymorphic fractions, respectively.

After correction for silent substitutions, 2 to 4 percent nucleotide substitutions in translated DNA would yield 5 to 9 percent amino acid differences. An electrophoretic study of 12 enzyme systems in *S. intermedius* has given a heterozygosity estimate of 0.18, which is not very different from the mean value for invertebrates (see Table 4). If we assume that $\overline{H} = 0.18$ corresponds approximately to one amino acid difference per five proteins, and that the average length of a protein is 300 amino acids, the electrophoretic data would reflect one substitution per 1,500 amino acids (Grula et al., 1982). The value obtained from the reassociation data is about 100 times greater (see above: 5 to 9 percent amino acid substitutions $\approx 1/15$). The difference may be due in part to the inability of detecting all amino acid substitutions by electrophoresis. But it seems likely that the larger proportion of the nucleotide diversity observed by reassociation involves DNA that does not code for amino acids. In any case, it deserves notice that the frequency of nucleotide heterozygosity observed by DNA hybridization (5 to 9 percent) is not very different from the value obtained by sequencing the $^A\gamma$ gene (2 percent).

DNA cleavage with restriction endonucleases is another method to estimate the proportion of nucleotide differences in the DNA. DNA-sequence polymorphisms have been detected by endonuclease digestion in human globin genes (Lawn et al., 1978; Tuan et al., 1979) and in the ovoalbumin gene of chicken (Lai et al., 1979). Jeffreys (1979) has examined in 60 unrelated human individuals a continuous DNA segment containing several globin genes of the beta family (i.e., most of the segment shown at the top of Figure 2). A cleavage site in one but not another DNA sequence means that the two sequences differ by at least one base pair at the site (each cleavage site contains four or more contiguous nucleotides). The number of cleaved sites is 52–54, amounting to 300–310 base pairs; the number of variant sites is three. The frequency of variable nucleotide sites may, then, be calculated as 3/300 = 1 percent. But this "intuitive" estimate can be shown to be biased; the corrected estimate is 0.5 percent (Ewens et al., 1981). Moreover, this is an estimate of polymorphism, not of heterozygosity. The latter can be estimated as 0.1 percent (Ewens et al., 1981).

The nucleotide heterozygosity value based on Jeffreys' data is about 20 times smaller than the value obtained from the actual sequence of the $^A\gamma$ gene. This may be accidental, e.g., due to the small number of nucleotides assayed by Jeffreys; or it may be that the endonuclease technique yields biased low estimates because restriction sites are more conserved than others. The second alternative may be questioned in view of the large frequency of nucleotide differences detected in the mitochondrial DNA of some but not other organisms by restriction endonucleases. In mice, for example, about 2 percent nucleotides are different between individuals in *Mus musculus* as well as in *M. domesticus,* whereas no differences have been detected in *M. molossinus* (Ferris et al., 1982). In primates, the frequency of nucleotide differences between individuals is 1.0 to 1.3 percent in chimpanzees, but only 0.3 percent in humans (Ferris et al., 1981*a*; see also Upholt and Dawid, 1977, for sheep; Avise et al., 1979*a*, for deer mice; Avise et al., 1979*b*, for gophers; and Brown and Simpson, 1981, for rats).

Although quantitative estimates of the amount of DNA-sequence variation cannot be provided with confidence for organisms in general, there can be no doubt that the variation is extensive. If the noncoding regions of genes are included, it seems likely that most, if not all, genes are heterozygous in every outbred individual. The amount of variation in the flanking sequences that occur between genes is also likely to be large.

VARIATION BETWEEN POPULATIONS

Environmental conditions vary from place to place and from time to time; the number of individuals in any population is limited. Hence, natural selection and random genetic drift affect gene frequencies, which will be different in different populations or at different times.

Genetic differentiation between local populations, even within distances of only a few kilometers, has been established with respect to such traits as growth habit and flower color in plants, color and pattern in snails and butterflies, chromosome arrangements in *Drosophila* and grasshoppers, etc. (Dobzhansky, 1970; Wright, 1978).

Studies of electrophoretic variation in *Drosophila* have shown that populations of the same species usually share the same most common allele, as well as most other alleles except those in very low frequencies. It is not, however, the case that populations are electrophoretically uniform. Even in *Drosophila,* nontrivial variation exists between local populations and even between seasons (Dobzhansky and Ayala, 1973). Table 10 gives two examples of local variation in *D. willistoni.* At the *Pgm-1* locus, allele *100* is the most common in all six island populations, but its frequency ranges from 0.70 to 0.98; correspondingly, the frequency of allele *104* rises from 0.01 in Carriacou to 0.30 in St. Lucia; local variation is even larger at the *To* locus. Local variation in some other organisms is considerably greater than in *Drosophila.* And it should be kept in mind that identity in electrophoretic mobilities does not necessarily imply identical allelic products (Chambers et al., 1981).

Local genetic variation can be summarized and quantified in a variety of ways. Genetic distance statistics have been used in a number of cases. Wright (1978) has used F-statistics in an extensive survey of electrophoretic data. F-statistics were originally developed to measure amounts of inbreeding and assortative mating but were later extended to evaluate interpopulational differentiation. F_{DT}, the statistic used here, may range from 0 to 1 and measures the amount of differentiation among demes relative to the maximum that would occur if the observed alleles were fixed in different populations. F_{DT} yields nonintuitive results and may be misleading, in certain extreme situations. If there are two or more alleles and all populations are fixed, $F_{DT} = 1$, independently of whether or not *some* populations are fixed for the same allele (and are, therefore, genetically identical). On the other hand, if populations are polymorphic, F_{DT} is always less than 1, even

TABLE 10. Allelic frequencies at two loci in island populations of *D. willistoni.* The number of genes sampled is about 200 for each locus in each locality. (After Ayala et al., 1971)

Gene: Allele	Grenada	Carriacou	Bequia	St. Vincent	St. Lucia	Martinique
To:						
86	0.05	0.21	0.54	0.56	0.14	0.19
100	0.94	0.79	0.46	0.44	0.85	0.81
Pgm-1:						
96	0.02	0.01	0.00	0.02	0.00	0.00
100	0.97	0.98	0.91	0.79	0.70	0.77
104	0.02	0.01	0.09	0.19	0.30	0.23

TABLE 11. Differentiation (F_{DT}) between local populations of species in which 15 or more gene loci have been studied by electrophoresis. (After Wright, 1978; modified)

Species	Loci sampled	Populations sampled	F_{DT}	PERCENTAGE OF LOCI WITHIN RANGE OF F SHOWN					
				0–04	.05–14	.15–24	.25–49	.50–74	.75–1
DROSOPHILA									
D. willistoni (S. America)	25	12	0.022	88	12	—	—	—	—
D. pseudoobscura	24	3	0.028	92	4	4	—	—	—
D. equinoxialis	23	10	0.029	87	13	—	—	—	—
D. willistoni (W. Indies)	24	6	0.041	88	8	4	—	—	—
D. obscura	30	3	0.067	90	7	3	—	—	—
D. pavani	16	14	0.126	56	31	13	—	—	—
D. paulistorum	17	17	0.255	47	18	18	6	12	—
HORSE-SHOE CRAB									
Limulus polyphemus	25	4	0.047	88	12	—	—	—	—
FISH									
Astyanax mexicanus (surface)	17	6	0.025	94	6	—	—	—	—
Astyanax mexicanus (cave)	17	3	0.650	59	12	—	12	6	12
REPTILES									
Anolis carolinensis (U.S.)	27	3	0.119	89	7	4	—	—	—
Sceloporus grammicus	20	4	0.444	65	5	5	10	10	5
Uta stansburiana	18	17	0.490	54	6	11	6	17	6
MAMMALS									
Sigmodon hispidus	23	6	0.067	87	13	—	—	—	—
Mus m. musculus	41	4	0.089	87	7	7	—	—	—
Peromyscus polionotus	32	31	0.394	50	19	9	9	9	3

when the populations share no alleles in common. But F_{DT} works fairly well for most actual situations observed.

Table 11 gives the value of F_{DT} for a number of organisms in which 15 or more loci have been studied by electrophoresis in at least three populations. The value of F_{DT} has been calculated first for each locus and then averaged for all loci. The table also gives the percent of loci falling in each value-range of F_{DT}. Not all loci behave similarly within a given organism. For example, in cave populations of *Astyanax mexicanus*, 59 percent of the loci are uniform ($F_{DT} = 0.00 - 0.04$) while 12 percent are nearly completely different in different populations ($F_{DT} > 0.75$). The largest degree of interpopulational variation in *Drosophila* occurs in *D. paulistorum*, but this is a special case because the populations surveyed belong to different incipient species. The value of F_{DT} is fairly large in *D. pavani*, but all other *Drosophila* species exhibit only moderate degrees of heterogeneity among populations, although the heterogeneity is far from trivial. Large differentiation between populations exists in one of the three species of mammals and appears to be the rule rather than the exception in lizards.

In summary, genetic polymorphism is not only extensive within a given population, but there exists in addition heterogeneity between local populations. Species do indeed possess enormous stores of genetic variation. The opportunity for evolutionary change is ample.

CHAPTER 5

THE MEANING OF PUNCTUATED EQUILIBRIUM AND ITS ROLE IN VALIDATING A HIERARCHICAL APPROACH TO MACROEVOLUTION

Stephen Jay Gould

DEFINITIONS

Punctuated equilibrium is a theory about the deployment of speciation in geological time (Figure 1). As such, it is about the tempo and mode of evolution. It holds, speaking of mode, that significant evolutionary change arises in coincidence with events of branching speciation, and not primarily through the *in toto* transformation of lineages (classical anagenesis). It maintains, speaking of tempo, that the proper geological scaling of speciation renders branching events as geologically instantaneous and that, following this rapid origin, most species fluctuate only mildly in morphology during a period of stasis that usually lasts for several million years (estimates for the average duration of marine invertebrate fossil species range from 5 to 11 million years—Raup, 1978; Stanley, 1979). Since "geologically instantaneous" is a fuzzy term, I make the operational suggestion that it be defined as 1 percent or less of later existence in stasis. This permits up to 100,000 years for the origin of a species with a subsequent life span of 10 million years, though I believe that most events of speciation occur much more rapidly.

As with most major issues in natural history, support for punctuated equilibrium relies on an argument about relative frequency, not a claim for exclusivity. I know nothing in evolutionary theory (nor can I envisage anything) that would render anagenetic transformation inconceivable a priori. Gradual phyletic transformation can and does occur. I claim merely that its relative frequency is low and that punctuated equilibrium is the predominant mode and tempo of evolutionary change. It is hard to put a number on "low" and "predominant." Would an off-the-cuff claim for 90 percent be sufficient to rescue me from a charge of winning my own argument by defining it for easy victory? If God appeared and informed us that 48.647 percent of significant evolutionary events had occurred via punctuated equilibrium, I would be more (but only slightly more) than half-satisfied. To be half-right, however, is a blessing in a complex world.

Two other definitional points need emphasis:

(1) Punctuated equilibrium is a specific claim about speciation and its deployment in geological time; it should not be used as a synonym for any theory of rapid evolutionary change at any scale. Since potshots at priority are a popular pasttime, some colleagues have claimed that all manner of men, from Darwin* to Simpson, said it all before. But these prior claims are either about true saltation in ecological time (DeVriesian macromutation, for example) or episodic anagenetic change—not about the consequences of ordinary speciation in geological scaling.

Simpson's important hypothesis of "quantum evolution," for example, has been cited as preemptive of punctuated equilibrium (Boucot, 1978). But quantum evolution is about rapid anagenetic change,

---

* The Darwinian claim is particularly egregious. No belief was more central to Darwin's thinking than gradualism; this may be the one point on which all Darwin scholars agree (Gruber, 1974; Mayr, 1978). Darwin's faith in gradualism was greater by far than his confidence in natural selection, though he often conflated the two—as in this clearly invalid statement: "If it could be demonstrated that any complex organ existed, which could not possibly have been formed by numerous, successive, slight modifications, my theory would absolutely break down" (1859, p. 89). To Darwin, the gradual continuity of change was part of the very definition of a natural process (Gruber, 1974, pp. 125–26). One may, of course, cite the single sentence insertion into the fourth edition of the *Origin,* in which Darwin allows that lineages undergo marked fluctuation in rates, with long periods of stability and short episodes of activity, but to what effect? No one ever believed in an absolute constancy of rates, and Darwin speaks here of anagenesis within lineages, not punctuated equilibrium. You cannot do history by selective quotation and search for qualifying footnotes. General tenor and historical impact are the proper criteria. Did his contemporaries or descendants ever read Darwin as a saltationist?

explicitly not about speciation and its consequences. In formulating his hypothesis, Simpson (1944) set it in contrast with two other modes of evolution: speciation and phyletic evolution. He also argued that speciation is unimportant in producing major evolutionary change: "This sort of differentiation," he wrote, "draws mainly on the store of pre-existing variability in the population. . . . The phenotypic differences involved in this mode of evolution are likely to be of a minor sort or degree. They are mostly shifting averages of color patterns and scale counts, small changes in sizes and proportions, and analogous modifications" (1944, p. 201). But punctuated equilibrium holds that accumulated speciation is the root of most major evolutionary change, and that what we have called anagenesis is usually no more than repeated cladogenesis (branching) filtered through the net of differential success at the species level.

Moreover, Simpson retreated on quantum evolution as the modern synthesis hardened around an adaptationist core (Gould, 1980). In 1953, he redefined quantum evolution as one end of the rate-continuum for phyletic evolution, or anagenesis—the very thing that punctuated equilibrium is not. Quantum evolution, he wrote, "is not a different sort of evolution from phyletic evolution, or even a distinctly different element of the total phylogenetic pattern. It is a special, more or less extreme and limiting case of phyletic evolution" (Simpson, 1953, p. 389). He also linked quantum evolution firmly to adaptation (he had denied this in 1944), arguing that such abrupt shifts imply an even more active control by selection: "Indeed the relatively rapid change in such a shift is more rigidly adaptive than are slower phases of phyletic change, for the direction and the rate of change result from strong selection pressure once the threshold is crossed" (1953, p. 391). By emphasizing an enlarged set of reasons for the differential origin and survival of species, punctuated equilibrium tries to move beyond the restrictive notion that evolutionary trends must reflect the direct advantages of phenotypes under regimes of natural selection (see later).

(2) Of the two claims of punctuated equilibrium—geologically rapid origins and subsequent stasis—the first has received most attention, but Eldredge and I have repeatedly emphasized that we regard the second as more important. We have, and not facetiously, taken as our motto: stasis is data. We do this for two reasons:

First, although all paleontologists have acknowledged the importance of stasis, they have traditionally not written about it, while the literal record of rapid origins has been widely discussed. (If evolution means change, then the absence of change is not worth talking about, or so the specious argument goes.) To be sure, this rapidity was dismissed as the artifact of an imperfect fossil record, but at least its literal appearance gained widespread attention. We found, however, that most "neontologists" (our slightly opprobrious term for all you folks who study modern organisms) were simply unaware of the prevalent stasis that affects so many species for millions of years.

Second, stasis (as data) is available for study. Rapid transitions are mostly recorded by an absence of information (though see Williamson, 1981, for an exceptionally fine-grained study of speciation events themselves). We argued that this absence does not record the missing millions of gradual anagenesis but the missing thousands of ordinary speciation. Still, absence is absence, and you can't do much with it. Stasis can be studied directly. The (potential) validation of punctuated equilibrium will rely primarily upon the documentation of stasis. Gaps can always be attributed to the traditional argument of imperfection and are therefore rarely decisive. It takes unusual stratigraphic resolution (sometimes possible of course) to prove that a morphological break records a true evolutionary punctuation, but stasis can be studied in conventional geological sections with their missing data of more than 90 percent (Schindel, 1980)—for 10 percent of 10 million years, widely dispersed over the interval, is a good sample.

We also believe that stasis is the more interesting phenomenon for evolutionary theorists. In our original formulation (1972), Eldredge and I argued that the geological scaling of punctuation is fully consistent with the most orthodox version of Mayrian allopatric speciation in peripheral isolates (see later). As such, the punctuations themselves need not record any unconventional evolutionary phenomenon. But pronounced stasis, as the usual fate of most species, was neither predicted nor expected by traditional Darwinism (Stebbins and Ayala, 1981). It is, therefore, an "interesting" phenomenon—as John Maynard Smith stated at the Chicago macroevolution meeting (October, 1981).

The initial response of Darwinians has been to attribute such stasis to stabilizing selection (Maynard Smith, personal communication; Charlesworth et al., 1982). If so, then it is a conventional phenomenon after all. I confess to great difficulty in accepting such an explanation. Stasis usually persists through millions of years and through the extensive climatic changes that any local area undergoes during such a long interval. I cannot believe that directional selection would be so inactive in such a context if it could, in the anagenetic mode, readily convert within-population variation to the larger scale differences that separate species and genera. I suspect instead that directional selection is at least sporadically active, and that it accounts for the mild and (at larger scale) directionless fluctuations that affect lineages in stasis. I would look instead to the panoply of suggestions—now so

86

imperfectly formulated, but all responsive to what I regard as a correct intuition about the stability of form (Webster and Goodwin, MS)—that would link stasis to a recalcitrance of inherited genetic and developmental programs, not to an absence of "push" from external sources: closed vs. open genetic systems (Carson, 1975); structural vs. regulatory genes (King and Wilson, 1975; Wilson et al., 1975); and developmental constraints (Alberch, 1980). All these proposals, with their emphasis on the importance of "internal" factors, do not lie within the spirit of modern Darwinism.

THE RELATIONSHIP OF PUNCTUATED EQUILIBRIUM TO NEONTOLOGICAL THEORIES OF SPECIATION

This section can be short, for there is little relationship or, rather, little constraint imposed by the existence of punctuated equilibrium upon modes of speciation. Speciation and its deployment over millions of years are events of different scale. By my one-percent criterion (see above), tens of thousands of years are usually available for the origin of full reproductive isolation.* Since times of this order are sufficient for almost any model of speciation (except "dumb-bell" allopatry, where a population splits into two roughly equal parts and each diverges slowly thereafter), punctuated equilibrium per se does not suggest or specify any mode. In fact, Eldredge and I wrote our first paper (1972) to argue that the most orthodox form of Mayr's "peripheral

---

* Reproductive isolation and the morphological gaps that define species for paleontologists are not equivalent. Punctuated equilibrium requires either that most morphological change arise in coincidence with speciation itself, or that the morphological adaptations made possible by reproductive isolation arise rapidly thereafter. Under classical extrapolationist ideas, speciation was little more than geographic variation extended, and reproductive isolation arose as the accidental by-product of sufficient adaptive divergence—thus suggesting a tight correlation between the genetics and morphology of speciation. Under several newer theories of speciation (White, 1978; Templeton, 1980*b*), reproductive isolation may arise rapidly, as an initial step that sets a potential (by producing independent populations) for the later divergent accumulation of adaptations. These newer ideas are a two-edged sword for punctuated equilibrium. Their emphasis on rapidity is favorable, but their decoupling of morphology and reproductive isolation raises an important question. If the morphological adjustments that produce a new stable system occur rapidly following reproductive isolation, then all is well. (I would apply the same one-percent criterion for these morphological changes; the reproductive isolation itself, under models that are virtually saltational in ecological time, is immeasurable in geological time). If morphological adaptations usually accumulate gradually (in geological perspective) with no tendency to any rapid initial setting and stabilization, then punctuated equilibrium is wrong. I tend to view species as stable systems, rapidly established in periods of instability and resistant to essential changes (though subject to all manner of minor fluctuation) thereafter. I believe that the fossil record of stasis supports this claim.

87

isolate" model (1963) implies punctuated equilibrium in geological scaling. In this light, I have been vexed that several colleagues have attacked punctuated equilibrium by casting doubt upon rapid parapatric and sympatric modes of speciation. I shall, of course, be pleased insofar as these modes are established at reasonable or high relative frequencies—the faster the better for punctuated equilibrium. But since punctuated equilibrium is consistent with the standard Darwinian model of the modern synthesis (Mayr, 1963)—indeed it was developed in this context (for these rapid modes either had not been proposed or had few supporters when we first wrote in 1972)—its vindication is not dependent upon the outcome of current debates about speciation. This does not mean, as other colleagues have charged, that punctuated equilibrium is soft, squishy, or unfalsifiable because most common modes of speciation are consistent with it. It only means that punctuated equilibrium is a phenomenon of different scale, and that its support or falsification will be tied to events at geological dimensions (including the relative frequency of stasis, the persistence of ancestors after punctuations, and the efficacy of species selection in producing evolutionary trends).

THE RELATIONSHIP OF PUNCTUATED EQUILIBRIUM TO MACROMUTATION

Punctuated equilibrium is not a theory of macromutation (Lande, 1980, notwithstanding); it is not a theory of any genetic process (though its geological geometry, if vindicated, exerts some constraint upon genetic modes—gradual and sequential allelic substitution will not be a good model for the origin of higher taxa if stasis be prevalent). It is a theory about larger-scale patterns—the geometry of speciation in geological time. As with ecologically rapid modes of speciation, punctuated equilibrium welcomes macromutation as a source for the initiation of species: the faster the better. But punctuated equilibrium clearly does not require or imply macromutation, since it was formulated as the expected geological consequence of Mayrian allopatry (Eldredge and Gould, 1972).

I do feel that certain forms of macromutational theory are legitimate, and I have supported them (Gould, 1980), though not in the context of punctuated equilibrium (I do have other interests, after all). I doubt that the initiation of species by macromutation has a high relative frequency, but even rare occurrence may produce important evolutionary results because major morphological shifts are themselves so uncommon. Illegitimate forms of macromutation include the sudden origin of new species with all their multifarious adaptations intact *ab initio,* and origin by drastic and sudden reorganization of entire genomes. Legitimate forms include the saltatory origin of key features (around which subsequent adaptations may be molded) and marked phenotypic shifts caused by small genetic changes that affect rates of development in early ontogeny with cascading effects thereafter.

Few evolutionists recognize (and Goldschmidt's colorful and absolutistic phrases didn't help his case) that Goldschmidt set his "hopeful monster" theory in this legitimate context. Goldschmidt was primarily interested in how development constrains and facilitates macroevolution. He defined hopeful monsters as phenotypic products of small genetic changes that impact early ontogeny. Cascading effects arise from potential alternative pathways of development already contained in inherited norms of reaction (as demonstrated in teratologies, phenocopies, mutants, and simple sexual differences—see my introduction to the republication of Goldschmidt's *Material Basis of Evolution,* Gould, 1982). Monsters may be hopeful because the regulative properties of development tend to channel perturbations along viable (if discontinuous) routes. (Most monsters, of course, are hopeless because they are not so regulated, and these play no role in evolution.) Since genetic differences between hopeful monsters and normal forms are minor, breeding may not be impaired (or heterozygote disadvantage will be small); under certain population structures (discrete small demes with inbreeding), small populations of homozygous hopeful monsters may be established (Goldschmidt, 1940, p. 207). The ultimate success of a hopeful monster depends upon the subsequent development of stabilizing adaptations around the saltatory key feature.*

Hugh Iltis' theory for the origin of the corn ear (Iowa meeting of the Society for the Study of Evolution, July 1, 1981) postulates a hopeful monster of pure (and legitimate) Goldschmidtian form. The ear evolves by trading off developmental pathways already present in teosinte (in this case, ordinary sexual differences). Iltis sees the central spike of the male tassel as the homolog of the female teosinte "ear."

---

* Lande (1978, 1980) has argued forcefully against hopeful monsters by elucidating the polygenic basis of morphological alternatives that Goldschmidt and others attributed to genetic saltation. Lande may well be usually right, and Goldschmidt almost always wrong, about the underlying genetic basis of such alternatives. From a morphologist's point of view, the important issue involves the alternative phenotypes themselves: is the phenotypic transition gradual and continuous, or saltational. Lande accepts that phenotypic transitions in highly polygenic traits can be saltational via threshold effects. In some genetic perspectives, the phenotypic style of transition may be epiphenomenal and the underlying genetic basis all important. But one may also argue (as Alberch, 1980, has forcefully done) that the phenotypic style represents an important issue in its own right, regardless of its genetic basis, and that the issue of abrupt vs. continuous phenotypic transition raises a host of interesting evolutionary questions at other levels: structural and developmental constraints and opportunities, in particular. As a paleontologist, I confess to a professional bias for viewing the phenotypic issue as fascinating *in se*; I emphatically do not deny to my geneticist colleagues their own preferences, and only plead for mutual acknowledgment of interest.

89

He regards the modern corn ear (female) not as the gradually enlarged teosinte ear (also female)—since teosinte ears are lateral to the first lateral branches—but as the saltationally transformed male central spike (which, as the modern corn ear, is terminal on the lateral branch). He argues that shortening of the lateral branch would bring the male tassel into a zone of feminization where, by apical dominance, it might enlarge itself suddenly and suppress ears below it. The whole process might be accomplished by no more new genetics than that involved in shortening the branch—given the "old" genetics of sexuality and development. Iltis even suggests that the branch might be shortened with no internal genetic change at all—various viruses and smuts may produce this effect. (Some teratologies of modern corn can produce female ears in place of male central spikes.) Human selection would preserve, propagate, and "improve" such a hopeful monster. I cannot judge Iltis' case, but, as a general argument, it is certainly legitimate in theory, interesting and worthy of pursuit. And just as certainly, our subtle gradualistic biases have impeded the formulation of such alternative propositions by directing attention to the gradual enlargement of teosinte ears.

These legitimate styles of macromutation are related to punctuated equilibrium only insofar as both represent different and unconnected examples of a general style of thinking that I have called punctuational (as opposed to gradualist or continuationist thought). I take it that no one would deny the constraining impact of gradualistic biases upon evolutionary theorizing. Punctuational thinking focusses upon the stability of structure, the difficulty of its transformation, and the idea of change as a rapid transition between stable states. Evolutionists are now discussing punctuational theories at several levels: for morphological shifts (legitimate macromutation), speciation (various theories for rapid attainment of reproductive isolation), and general morphological pattern in geological time (punctuated equilibrium). These are not logically interrelated, but manifestations of a style of thought that I regard as promising and, at least, expansive in its challenge to conventional ideas. Any manifestation may be true or false, or of high or low relative frequency, without affecting the prospects of any other. I do commend the general style of thought (now becoming popular in other disciplines as well) as a fruitful source for hypotheses.

PUNCTUATED EQUILIBRIUM AND SPECIES SELECTION: HIGHER LEVELS AND EVOLUTIONARY TRENDS

If the geometry of punctuated equilibrium prevails within the history of most clades (Figure 1A), then traditional arguments for evolutionary trends, the most important phenomenon studied by paleontologists, must fail.

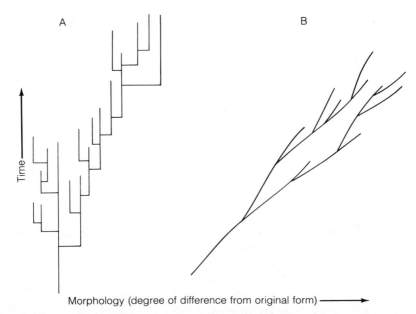

FIGURE 1. An evolutionary trend under punctuated equilibria (A) and phyletic gradualism (B). Speciation occurs in the phyletic model, but it neither changes the direction of a trend nor accelerates its rate and therefore does not power the trend. Differential success of species produces the trend under punctuated equilibrium.

In the traditional argument, trends are products of anagenetic transformation within lineages, mediated by natural selection.* Of course, no one has ever held that sustained trends occur in the absence of some speciation within clades. The standard phyletic model does not deny speciation, but simply grants it no role in directing the trend (Figure 1B). Since speciation entails no change in direction or acceleration of rate in phyletic models (Figure 1B), it does not enhance the directional component. Its role, instead, is to iterate a favorable (and phyletically generated) adaptation into several lineages, thus aiding its spread and compensating for any extinction. (After a split, ordinary phyletic evolution can occur in two lineages rather than just one.) It

* Curiously enough, through more than a century's debate among paleontologists about the causes of trends, one feature has endured: the assumption that trends occur in the phyletic mode. The agent of phyletic transformation has been hotly contested (with nonadaptive "internalist" theories like orthogenesis and racial life cycles, or adaptive internalist theories like Lamarckism, favored until about 1940, and natural selection triumphant thereafter). Through all this intense discussion, no serious challenge was offered to the inarticulated major premise (hidden assumption) of the argument itself: the phyletic mode.

91

can also enhance opportunities for further change by producing several units that can respond in different and independent ways to diverse environmental factors. This picture lies behind the traditional separation of cladogenesis from anagenesis, with their opposed significances as generators of "diversity" and "progress" respectively. Simpson had this picture in mind when he deemphasized speciation as an agent of major evolutionary change.

If we explore the speculative explanations offered for such classical trends as complexification of ammonite suture lines (Schindewolf, 1950) or increasing symmetry of the cup in Paleozoic crinoids (Moore and Laudon, 1943), we find that they presuppose direct advantages for morphology under phyletic regimes of natural selection (strengthening of the ammonite shell, or perfection of radial symmetry to permit collection of food from all directions in sessile organisms, for example). We also find that such explanations have been notoriously unsuccessful—a plethora of contradictory speculations that never bring resolution. Perhaps the problem lies deeper than our failure to devise a good adaptive story; perhaps we must challenge the basic (and usually both unstated and unrecognized) assumption that proper explanation must lie with such a story, if only we could find the right one.

Under the alternate geometry of punctuated equilibrium (Figure 1A), there is no phyletic component to direct a trend; species arise (in geological time) with their differences established at the start, and do not change substantially thereafter. Trends must therefore be the product of a higher-order sorting that operates via the differential birth and death of species considered as entities (the same role that individual organisms, which do not change evolutionarily during their life, play in microevolution). This higher order sorting of species, produced by differential origin and extinction (Figure 2), must direct evolutionary trends within clades (macroevolution) just as natural selection, acting by differential birth and death of bodies, directs evolutionary change within populations (microevolution).

What then shall we call this higher order sorting? Since 1975, when Stanley introduced the term, it has generally been designated as "species selection". (After some initial reluctance—for the wrong reasons—Eldredge and I have used species selection in this sense as well). In a purely descriptive sense, such a usage is amply justified. If species are stable and long lived, then trends may be described as a result of their differential success as entities. This descriptive meaning also provides considerable utility, for it suggests new procedures in analyzing evolutionary patterns (as Stanley, 1979, has exploited so well in tabulating longevities and rates of origin and extinction of species, instead of inferred rates of transformation).

Such descriptive usages are also well established in parallel cases. Provine (personal communication, and in preparation) shows that Se-

92

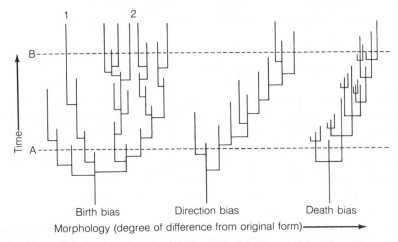

FIGURE 2. Explanations for evolutionary trends under punctuated equilibrium. Left: birth bias. We begin at time A with two kinds of species, each of equal number within the clade. At time B, the descendants of both kinds remain at the ancestral mode, but differential speciation permits one kind to dominate the clade. Center: direction bias. Speciation occurs more often to the right than to the left. Right: death bias. Most species give rise to two descendants, one in each direction. But species to the right live longer than species to the left, thus helping to power the trend. Birth and death biases are modes of species sorting; origin bias is the macroevolutionary analog of mutation pressure.

wall Wright has often invoked "interdemic selection" as an economical and heuristic description when the differential success of certain demes arises entirely from the competitive success of their constituent individuals.

Although these descriptive uses have respectable pedigrees, they are subject to an unfortunate and crucial confusion in their intersection with another issue to which they are intimately related—the "units of selection" controversy. Terms like "species selection" and "interdemic selection" are so similar to others well understood as expressions of selection based upon group-level properties—"group selection" and "family selection," for example—that the broad descriptive sense will inevitably be misinterpreted as a narrower claim for true group effects. This confusion is enhanced by the fact that an important subset of cases within both interdemic and species selection (broad sense) do involve true group selection (narrow sense).

Eldredge and I, and Stanley, have often emphasized that many

93

trends properly described as a sorting out of species (species selection in the broad sense) may be powered by traditional natural selection. (The obvious cases involve trends caused by differential longevity of species, when increased duration reflects the competitive success of individuals).

Nonetheless, the inevitable confusion between broad and narrow senses—especially the assumption that claims for group selection are always involved—is most unfortunate, especially since the existence of true group selection in *some* (but not all) trends is an important component of our argument for the independence of macroevolution. I therefore regret that we defined "species selection" in the broad sense, for this usage guarantees persisting confusion. I strongly recommend that the term "species selection" be confined to the narrow sense of true group selection—claims for selection among species based on species-level properties. I shall present an example of true species selection (p. 95), argue for its high relative frequency in the generation of macroevolutionary trends, and claim that it is a dominant process when trends are produced by differential origination rather than differential extinction.

The key issue for the independence of macroevolution is not whether species selection operates in all trends (it does not), but whether the necessity, under punctuated equilibrium, of regarding trends as a higher-level sorting of species implies a new level in a hierarchy of evolutionary explanation.

The recognition of new levels does not require group selection at that level in all cases. It does require the emergence of properties unpredictable from the behavior of entities at the next lower level. Hierarchy depends upon individuation; each level must be based upon entities sufficiently stable to acquire evolutionary properties of their own. Thus, Wright's shifting balance theory requires deme structure, but not interdemic selection (narrow sense) in all cases. (Interdemic selection does often occur, as when the emigration of propagules that invade other demes is regulated by such demic properties as population size). For example, in panmictic populations, genetic drift can only act as an agent of direct evolutionary change, whereas it functions as a mechanism for the production of variability at the deme level in Wright's theory (by driving demes into valleys and permitting selection to draw them up to adjacent peaks). Thus, deme structure produces a new and emergent role for genetic drift, and demes are a legitimate evolutionary level above individual bodies.

It is in this sense that punctuated equilibrium is crucial to the independence of macroevolution—for it embodies the claim that species are legitimate individuals, and therefore capable of displaying irreducible properties. Species selection is the strongest argument for macroevolution; but simple individuation is enough. A trend arising from differential species longevity, and due wholly to the success of competitively superior individuals in natural selection, is very different from a trend produced by the anagenetic transformation of a lineage. We must study different things—species longevities, rather than rates of transformation within species, for example. The rapid origin and subsequent stability of species guarantees the emergence of properties arising from the legitimate status of species as individuals.

The recognition that punctuated equilibrium implies a new hierarchical level by treating species as legitimate individuals has two important implications for evolutionary theory. First, it expands markedly the classes of hypotheses available for explaining evolutionary trends. Previously, an unexamined assumption of phyletic transformation required that morphological trends be viewed as direct products of natural selection acting upon advantages of individual organisms (stronger shells, better access to food, etc.). But if trends record the differential success of species, then a set of previously unexplored explanations must be considered.

As stated above, conventional natural selection remains as an important cause of trends, since advantages to individuals may produce differential longevity for species (trends to increasing brain size in mammalian lineages, perhaps). But other possibilities arise. Trends may occur simply because some kinds of species speciate more often than others, not because the morphologies so produced have any advantages under natural selection (indeed, such a trend will occur under random extinction). Or some species may survive longer than others because they inhabit a certain kind of environment, not because their morphologies are "better" in any conventional sense.

To illustrate these newer styles with the example most widely discussed by paleontologists of late (Vrba, 1980; Hansen, 1978, 1980; Jablonski, 1980), many clades of marine invertebrates exhibit a trend towards increased frequency of stenotopic species, often leading to the elimination of eurytopes completely, as in the volutid neogastropods (Hansen, 1978). (Stenotopes are narrowly adapted to definite environmental factors; eurytopes can tolerate a broad range of environments.) Tabulations of origination and persistence of species suggest that the trend is powered by a markedly greater rate of speciation among stenotopic species (there may also be a "conversion bias" of eurytopes to stenotopes—see later). In fact, the trend is often maintained even in the face of higher extinction rates in stenotopes as well (the birth differential exceeds the death differential and the trend continues). These higher rates of origination may be the result of greater possi-

95

bilities for the isolation of small populations in species that brood their young (as marine stenotopes generally do—eurytopes tend to have planktonic larvae, hence too much gene flow to permit isolation according to the argument above). But the morphological differences that mark the trend and allow paleontologists to recognize it (size of the protoconch, or embryonic shell, for example) may confer no advantages in terms of natural selection. Vrba (1980) recognized this unconventional property of morphological trends when she coined her "effect hypothesis." The trend to smaller protoconchs in clades that differentially accumulate stenotopic species (see Shuto, 1974) does not record natural selection acting upon protoconchs. The trend is merely an effect of differential speciation. Vrba shows a picture of justice blindfolded to illustrate her point. And it is well taken, for the world of directed morphological change may be far more capricious, or rather (and literally) coincidental, than we have imagined.

The other implication—species selection as a common phenomenon—raises a larger issue about the independence of macroevolution, and brings us face to face with the old and vexatious issue of units of selection (Lewontin, 1970; Wimsatt, 1980; Sober, in press). The Darwinian research program emphasizes reduction to the level of individual selection (as the strong reaction against Wynne-Edwards' hypothesis of group selection for the evolution of altruism illustrates—Williams, 1966). [Some hyperdarwinians (e.g., Dawkins, 1976) have even tried to transgress the sanctity of individual bodies and take their reduction of all higher-order phenomena right down to the level of the gene itself.] Sewall Wright told me (interview, October 1980) that he regarded "overemphasis on individual selection" as the most restrictive and unfortunate feature of the Darwinian modern synthesis (see Bock, 1979, for a very forthright admission and support of this reductionistic character of modern Darwinism). Despite strong arguments for true group effects (Wade, 1977, for example), and some convincing examples (as the case of the t allele in mice—see Dunn, 1956), evolutionists are still reluctant to acknowledge hierarchy with legitimate selection occurring among groups at several levels (demes, species, clades) and still feel more comfortable with reduction to individual selection. How else can one explain the delight of so many evolutionists with the theory of kin selection. It is a powerful and important theory to be sure, and I do not doubt its veracity and range, but we must also acknowledge its intellectual function within the Darwinian paradigm: It is an argument that saves individual selection in the face of phenomena that would appear, prima facie, to exclude it—for individuals can act in their own best interests by favoring kin over personal survival or reproduction.

Most trends that are powered by differential extinction may require no more than traditional individual selection—as when enhanced survival of a species arises from the competitive success of its individuals. But trends that are powered by differential origin must often include an important component of species selection. Propensity to speciate is not generally a property of individuals, since it depends so crucially on population sizes, density of habitation, and rates of migration related to density and size. Returning to our previous example: if stenotopic species take over a clade by differential speciation, then what role does natural selection play? In stenotopic species, all females brood their larvae, and we may find no genetic variance at all for this life history vs. planktotrophy. Moreover, stenotopic and eurytopic species within the clade probably do not compete directly and may not even exploit the same resources. How then can natural selection, acting on individuals, favor the increase of stenotopy within the clade? I think it would be an unwarranted extension of our usual meaning of natural selection to argue that the role of a trait in parcelling itself into several taxa thousands of years hence (with no reference to benefits of the moment) can constitute its selective advantage. Moreover, an individual's success in contributing to speciation is critically dependent upon the genetic make-up of other individuals in the species. A single stenotope in a population of eurytopes has little opportunity to establish a species through the isolation of its progeny (too low a probability for founding an isolate and too much opportunity for its destruction, if founded, by gene flow from planktotrophic conspecifics). The same stenotope in a population of stenotopes has a greater chance (more conspecifics to produce offspring for the isolate, and less chance of its amalgamation with the parental population). Since the *same* genotype has different fitness for speciation as a consequence of group membership, we encounter an irreducible group effect. (I thank Eliot Sober for helping me to work through this argument, though I am not sure that he agrees with it.)

A NOTE ON HIERARCHY AND EMPIRICS

The need for hierarchy in evolutionary theory is a contingent fact of the empirical world, not a mere issue of semantics or methodological styles. We can construct a possible and nonhierarchical world in our thoughts; indeed, evolutionary theory has operated for many years under the assumption that such a world exists. The extrapolationist, phyletic model requires no hierarchy; if it reflected our world, reductionism would triumph. Punctuated equilibrium is the empirical claim that invalidates such a model by treating species as entities. This is its major importance in the structure of evolutionary theory.

97

The issue is larger than the independence of macroevolution. It is not just macroevolution vs. microevolution, but the question of whether evolutionary theory itself must be reformulated as a hierarchical structure with several levels—of which macroevolution is but one—bound together by extensive feedback to be sure, but each with a legitimate independence (see Eldredge and Cracraft, 1980). Genes, bodies, demes, species, and clades are all legitimate individuals in some situations (see Ghiselin, 1974, and Hull, 1980, on species as individuals), and our linguistic habit of equating individuals with bodies is a convention only. Each type of individual can be a unit of selection in its own right. Natural selection operating on bodies will not encompass all of evolution. Genes are units of selection in the hypothesis of "selfish DNA" (Doolittle and Sapienza, 1980; Orgel and Crick, 1980); demes are units in Sewall Wright's shifting balance theory (1968-1978). Species represent one level among many; evolutionary theory needs this expansion.

INTERMEDIATE LEVELS AND SHIFTING BALANCE: A FRIENDLY CHALLENGE TO PUNCTUATED EQUILIBRIUM

When Eldredge and I wrote our initial paper (1972) we did not understand Sewall Wright's theory of shifting balance, as a result both of our own shortcomings and of the general eclipse (now happily reversed) that Wright's work suffered at the time. Hence we relied on the supposed resistance of large populations to change and drew our contrast (in the standard Mayrian model) between small and fecund peripheral isolates and large central populations inertially restrained by the homogenizing force of gene flow. In this view, evolutionary change is associated with speciation—and we therefore made this equation in formulating punctuated equilibrium.

This argument is strong if we accept natural selection acting upon bodies as the primary agent of evolution (as Eldredge and I did at the time, however unconsciously). With too many organisms and too much panmixia, rare and favorable variations cannot readily spread and accumulate. In other words, we were caught in the very conventionality we have since tried to oppose: an overreliance upon individuals and natural selection.

Sewall Wright has long recognized that, under his shifting balance theory, demes can act as units of selection; originally, he referred to his idea as the "two-level theory," for natural and interdemic selection (Wright, personal communication). If demes act as an intermediate unit of selection between individuals and species, then the conceptual apparatus of inertial central populations vs. small and changeable isolates loses its justification in species with appropriate division into a large enough number of sufficiently independent demes to fuel the process of interdemic selection. Evolution can be just as effective in large populations as in small, perhaps even more so in large populations that support many demes. There is nothing special about speciation as an agent for substantial evolutionary change; the phyletic mode, operating by *both* interdemic and natural selection (not by natural selection alone) becomes effective.

Wright (in press) has challenged punctuated equilibrium on this basis, arguing that the general mode of change we call "punctuational" can be encompassed within the phyletic mode under the shifting balance theory, and not only via speciation, as in punctuated equilibrium. I regard this statement as an important challenge to punctuated equilibrium, but a welcome and "friendly" one for two reasons. First, it affirms the importance of punctuational change in general, while proposing a mechanism for it different from punctuated equilibrium (Wright has long argued that higher taxa arise rapidly when ecological opportunities permit species to move from the static to the dynamic phase of shifting balance—see Templeton, Chapter 2). Second, it is a challenge rooted in the notion of levels above natural selection operating on individuals—in this case, interdemic selection permitting rapid transitions in the phyletic mode. The vindication of hierarchy as a challenge to the reductionistic character of traditional Darwinism is a far more important issue than the relative importance of any particular higher level. We have emphasized the level of species, Wright that of demes.

In fact, Eldredge and I (Gould and Eldredge, 1977, pp. 141–145) did recognize the potential importance of intermediate levels between individuals and species in arguing that punctuational sorting of clones in asexual forms would translate to gradual change at the "species" level (see Figure 3). But, in our ignorance of Wright's work, we viewed this intermediate level as a phenomenon peculiar to asexual forms and did not appreciate the generality of the argument.

While I appreciate the logic of Wright's argument, I will still place my money on the greater relative importance of punctuated equilibrium in the general pattern of evolution within clades. First, since shifting balance yields either gradual or punctuational change within established species, while punctuated equilibrium champions the vastly greater relative frequency of stasis, my opinion that stasis is empirically so predominant leads me to favor punctuated equilibrium. Second, shifting balance is crucially dependent upon the subdivision of a species into enough sufficiently independent demes, and I am not confident that a large fraction of the world's species are so constituted.

In any case, shifting balance and punctuated equilibrium yield

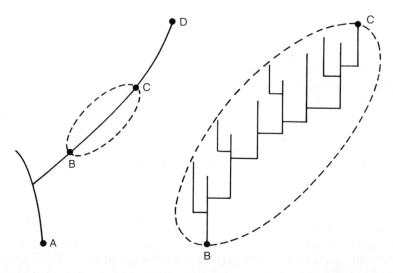

FIGURE 3. Punctuation at an intermediate level can yield gradual change at the species level (from Gould and Eldredge, 1977, p. 142). A punctuational pattern among clones of an asexual species (right) yields continuous change in the origin of a new species (D) from its ancestral lineage (A).

different predictions about the pattern of evolution, and can be tested in at least two ways in the fossil record. First, the relative frequency of stasis within species will be indicative, though not decisive: high for punctuated equilibrium (I ventured 90 percent at the beginning of this chapter), lower for shifting balance. Second, and more important, since punctuational events can occur in the phyletic mode under shifting balance, but by branching speciation under punctuated equilibrium, the persistence of ancestors following the abrupt appearance of a descendant is the surest sign of punctuated equilibrium. The literature on punctuational transitions in the fossil record includes both patterns (see Hallam, 1978, and Kellogg, 1975, for rapid transitions within lineages and Williamson, 1981, and the original examples in Eldredge and Gould, 1972, for persistence of ancestors), but we have as yet insufficient evidence to establish a clear relative frequency in any major group. I hope that this line of research will be exploited.

The existence of these alternative explanations for punctuational change suggests an interesting subject for evolutionary thought. Do different taxonomic groups display varying characteristic styles, and do these variations reflect different population structures or habitats (a fruitful point of potential contact between micro- and macroevolutionary studies). Eldredge and I are invertebrate paleontologists, and the fossil record (as many neontologists do not appreciate) is predominantly the story of marine invertebrates. Perhaps these are the very species least likely to exhibit a deme structure appropriate for the operation of shifting balance (enhanced gene flow and less environmental heterogeneity in the marine realm); and perhaps the predominance of punctuated equilibrium in the fossil record reflects this bias of its composition, thus rendering punctuated equilibrium less general than we have supposed. We (Gould and Eldredge, 1977, p. 141) have already identified the intermediate level of sorting among clones as a reason for potentially higher frequencies of gradualism at the species level in marine microfossils. The intermediate level of interdemic selection may produce the same effect for those species so structured that shifting balance can be effective. I hope that this too will strike the interest of students and become a subject for research.

A MAJOR INADEQUACY IN THE INITIAL FORMULATION OF PUNCTUATED EQUILIBRIUM: THE ROLE OF DIFFERENTIAL SPECIATION

Natural selection operates either by differential death or differential birth. All evolutionists appreciate this of course, but a cultural legacy long antedating Darwin has led us to emphasize death and the literal struggle for existence: nature red in tooth and claw, in Tennyson's old cliché. We still view evolutionary change by differential birth as an odd process, occurring here and there to be sure (notably in regimes of r selection), but not contributing to "progress" or increasing complexity.

As Gilinsky (1981) has demonstrated, both Eldredge and I (1972, 1977) and Stanley (1975, 1979) followed this restrictive legacy in our original formulations. We acknowledged the analog of differential birth (speciation) in our words, but formulated our discussions almost entirely in terms of differential death (extinction). The oft-repeated statement that speciation may be random with respect to the direction of evolutionary trends reflects the tight link that we made between the sorting of species and differential extinction.

I now believe that this formulation was inadequate in its misplaced emphasis (as Gilinsky, 1981, argues), and that evolutionary trends powered by differential birth are both common and more interesting in their unconventional implications. A trend can occur via differential speciation in two ways, only one of which may represent species selection (see Figure 2). In "birth bias" (Figure 2, left), a clade contains two or more kinds of species (analogs of alleles) at an initial time A, and the increased representation of one kind at a later time B arises only from its higher speciation rate (probability of extinction across

101

all kinds may be constant). Trends are a product of differential origin. In the example given earlier in this chapter, stenotopes may increase in frequency and take over clades because their rates of speciation are higher.

In "direction bias" (Figure 2, center, and Stanley, 1979), speciation is more likely to occur in one direction than in others, and trends may also arise in the face of purely random extinction. Species *selection* cannot be invoked here, for we have encountered the higher-level analog of mutation pressure, or differential direction of variation itself. Such direction bias may be most common in cases of "ontogenetic channeling," in which size increase (sufficiently common for codification as Cope's Rule) leads to a prolongation of ontogenetic allometries by hypermorphosis (Gould, 1977). Ammonite suture lines may become more elaborate by such a mechanism since increased complexity is a virtually universal pattern in ontogeny. Direction bias may also play a role in evolutionary trends towards stenotopy for a different reason. As R. Strathmann argues (personal communication), planktotrophic (eurytopic) larvae tend to build complex structures (ciliary bands primarily) that enable them to stay afloat. Benthic larvae (brooded stenotopes) lose these structures. As Dollo stated long ago, complex structures that are truly lost in evolution do not reevolve in the same manner (see Gould, 1970). Thus, we have a strong "conversion bias." Planktotrophs can lose their ciliary bands and become benthic, but the simpler larvae of stenotopes cannot rebuild the bands and float again. Speciation may run along a one-way street in this case.

I argued earlier in this chapter that trends powered by differential extinction may often be attributed to conventional natural selection operating upon bodies. Differential origin rarely fits into the strict Darwinian mold for two reasons. First, birth bias often requires that we consider species as irreducible units of selection. Differential origin is the primary realm of species selection. Second, we encounter a different non-Darwinian factor in direction bias. I see no reason to invoke group effects here; ciliary bands may be lost and ammonite sutures may become more complex via the usual route of natural selection. But speciation in macroevolution is the analog of birth in microevolution. Therefore, if new species arise with morphologies predictably different from their parents, evolutionary trends can be powered by a bias in the raw material itself, an analog to mutation pressure in microevolution. But randomness of raw material is a primary ingredient in Darwinian explanations. In short, the two processes of differential origin produce non-Darwinian effects in utilizing species as units of selection (birth bias), and in establishing as important a bias in directional production of raw material, the analog of mutation pressure (direction bias).

These cases of differential origin also illustrate important differences in style and emphasis between micro- and macroevolution. In microevolution, so many individuals (bodies) are produced with so little average difference among them that both differential birth of bodies with a favored allele and differential conversion to that allele by mutation pressure can usually be swamped by natural selection acting via the death of bodies. In this sense, our conventional feeling about the power of natural selection is contingent upon the existence of copious and similar individuals within a bounded population. In macroevolution, we have far fewer individuals (species) within our bounded unit (clade), and average differences among individuals are far greater. In this situation, selection upon death (extinction) may not be able to overcome either a birth or a direction bias, and much of the "creativity" of macroevolution may lie in the generation of variation (speciation) itself, not in the differential culling of this variation. This role for differential origin also removes the most obvious objection to the importance of species selection in evolution: Since so many individuals (bodies) exist at the level of populations and so few (species) at the level of clades, one might argue that individual selection must always swamp species selection. If trends arose primarily through minor differences in longevity, this argument would be strong, but birth biases in a small population of species will rapidly generate trends under random extinction.

CONCLUSION

The differences and similarities between macro- and microevolution have been a contentious issue within evolutionary theory from the very start. The absolute nature of such a separation lies at the basis of our first coherent system, that of Lamarck. Lamarck contrasted progress up the ladder of complexity, mediated by an internal "force that tends incessantly to complicate organization," with tangential adaptations induced by the external "influence of circumstances," transferred to heredity by the inheritance of acquired characters. This tradition of an absolute separation persisted throughout the nineteenth century in such macroevolutionary theories as orthogenesis and racial life cycles, in which foreordination of end results renders natural selection of the moment irrelevant to any larger trend. Goldschmidt (1940) revived this tradition in our century when he argued for a complete separation between geographic variation and speciation, and for a different genetic style (material basis) for the origin of species and higher taxa (Gould, 1982).

Perhaps Darwin's greatest achievement lay in counteracting this

103

tradition (he did not invent evolution itself after all) with a strong argument for continuity and reduction of macroevolution to natural selection acting upon individual bodies (see Ruse, 1980). This argument provided a powerful heuristic. For the first time, all the data that we can see and manipulate in our time scale (artificial selection, allelic change in natural populations) were no longer an epiphenomenon to a grand and mysterious cosmic directionality, but the stuff of evolution itself.

How can we mediate between these two positions? We do not wish to advocate the despairing conclusion that micro- and macroevolution are absolutely separate in principle and that nothing about one illuminates the other (the Lamarck-Goldschmidt tradition). But shall we accept Darwinian continuationism at the price of ignoring the legitimately independent properties of levels higher than natural selection operating upon organisms? Must the choice be either despair and acknowledgment or ignorance and bliss?

I believe that the concept of hierarchy resolves this pseudoproblem by incorporating the favorable features of both positions. Levels do not achieve their independence because some fundamentally new genetic process emerges at their scale. Rather, the same processes of variation and selection operate throughout the hierarchy. But they work differently upon the varying materials (individuals) of ascending levels in a discontinuous hierarchy. We cannot learn everything we need to know about evolutionary trends by studying what happens within demes, if only because species can act as units of selection. Important ties of feedback unite all levels, but new modes emerge at higher levels and reduction to natural selection upon organisms will not render all of evolution. Nothing about microevolutionary population genetics, or any other aspect of microevolutionary theory, is wrong or inadequate at its level. Little of it is irrelevant to students of macroevolution. But it is not everything.

When a proper hierarchical theory is fully elaborated, it will not be entirely Darwinian in the strict sense of reduction to natural selection acting upon organisms. Yet I suspect that it will embody the essence of Darwinian argument in a more abstract and general form. We will have a series of levels with a source for the generation of variation and a mode (or set of modes) for selection among individuals at each level (Arnold and Fristrup, 1982). The superseding of strict Darwinism may establish the Darwinian style of argument in its most general form as the foundation for a truly synthetic theory of evolution.

CHAPTER 6

TOWARD A UNIFIED SELECTION THEORY

Roger Milkman

The main day-to-day effect of natural selection is the maintenance of the status quo, the stabilization of the phenotype. To a relatively small directional residue, we attribute the great panorama of evolution.

To understand the process of selection fully, and to be able to analyze a variety of cases of selection, we clearly need a unified selection theory. Several aspects of the subject need unifying, and I will list them. Then, to be successful, selection theory must reconcile the usual existence, in a species for a particular phenotype, of two things: first, an optimum value, and second, abundant variation with a genetic basis. I will present a three-part argument which results in such a reconciliation. And finally, I will address the conflicting evidence as to the adaptive and selective significance of allelic polymorphisms and offer a view that resolves the conflict.

WHAT MUST BE UNIFIED

There are three aspects of selection within which unification is required.

First, parameters relating to allele frequencies and parameters relating to phenotypic values must be made part of one system.

Second, the four basic types of genetic selection models must be analyzed in common terms.

Third, the effects of selection and the effects of sampling (random genetic drift) must be treated as collaborating influences on change and on stability.

105

Genetic and phenotypic parameters

Traditionally, there have been two approaches to selection, similar at first glance but separated by a deep gulf. One, used by population geneticists, is called the general selection model. It begins conceptually with the Hardy-Weinberg distribution of genotypes, which plays the same ideal role in population genetics that Newton's First Law of Motion does in mechanics, as Ayala and Kiger have pointed out (1980). That is, genotype frequencies are described in the absence of forces acting on them. Actually, genotype frequencies are altered by differential fitnesses, as well as by sampling, and the alleles in the successful gametes combine to produce, not necessarily at random, the genotypes of the next generation. The differential fitnesses must result from the impact of the genotypes on phenotypes, since it is the phenotype upon which selection operates, but the phenotype is merely implicit in this model. Thus, it is genotypically explicit and phenotypically implicit.

The other side of the coin is the phenotypically explicit, genotypically implicit model of quantitative genetics and biometrics. Here the particular phenotypic values of interest are measured, and the observation of a response to selection is understood to reflect the contribution of (additive) genetic variation to the phenotypic variation. *Which* genes are involved, and their particular properties, are neither necessary to the use of the model nor usually practical to investigate. Predictions regarding subsequent generations can still be made.

Both approaches model selection, and they share some terms as well, often with the same symbols. Thus it is possible to speak of fitness in terms of genotypes or of phenotypes, and variance in size or in allele frequency. Nevertheless, the central parameters of the two approaches have been effectively isolated from one another until very recently. These central parameters are, respectively, s (selection coefficient) for population genetics and i (selection differential) for quantitative genetics. While there have been limited demonstrations or suggestions of a relationship between s and i in the past (see Milkman, 1978a, for a brief review), a proof with considerable generality did not appear until recently (Milkman, 1978a, b). It was followed quickly with a proof of greater generality and elegance (Kimura and Crow, 1978). The relationship is a simple one

$$s_{ij} \approx i g_{ij}$$

where s_{ij} is the selection coefficient on genotype ij; i is the selection differential, to be defined explicitly below, and expressed in standard deviation units; and g_{ij} is the phenotypic effect of substituting genotype ij for a reference genotype. It is also expressed in σ units. For example, we may consider three genotypes, *AA, Aa,* and *aa,* in which *AA* has

106

the greatest fitness. We set this greatest fitness at 1, and the other fitnesses, w_{Aa} and w_{aa}, are expressed relative to w_{AA}. Thus, AA is the reference genotype: $s_{AA} = 1 - w_{AA} = 0$; $s_{Aa} = 1 - w_{Aa}$; and $s_{aa} = 1 - w_{aa}$. Since AA is the reference genotype, the phenotypic values associated with other genotypes are compared with that of AA. Thus g_{aa} represents the phenotypic difference that results from substituting aa for AA. Finally, the selection differential, i, is not concerned with genotypes. It is the difference between the population mean phenotypic value, $\bar{x}$, and the selected mean phenotypic value, $\bar{x}_s$, which is $\int x w_x f_x dx / \overline{w}_x$ or $\Sigma x_i w_i / n \overline{w}$ (the mean of the products of each individual's phenotypic value and its fitness). Since we are dealing in standard deviation units, $\bar{x} = 0$ by definition, and so i is numerically equal to $\bar{x}_s$ in this case.

This relationship is quite precise when g is small (≤ 0.02, for example), and it applies to all frequency distributions, continuous and discontinuous, and to modal (and multimodal) as well as to monotonic fitness functions. This last point means that stabilizing as well as directional selection is included; that the optimum phenotype may be of intermediate value. The dimensions of s and i are t^{-1}, whereas g is a dimensionless value. Figure 1 illustrates g, and Figure 2 shows how the bridge between s and i links the two major sets of selection-related parameters.

The assignment of a fitness value to a genotype such as AA is generally simple and often arbitrary. Since there are usually just a few genotypes under consideration, the set of their respective fitnesses is easy to keep in mind. Phenotypes, on the other hand, often involve a distribution of many values. Moreover, the distribution may be continuous. Here the simplest way to visualize the relationship be-

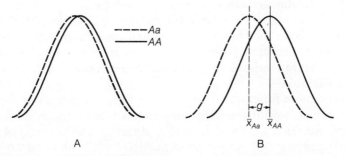

A   B

FIGURE 1. Distribution of phenotypic values for individuals of two A-locus genotypes (left), and exaggerated displacement (right) to illustrate g. (After Milkman, 1978a.)

107

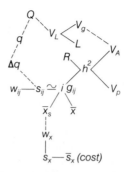

FIGURE 2. Interrelationships between some important parameters of population genetics and quantitative genetics. The variables i, g, L, q, s, V, and w are defined in the text; h^2, heritability; Q, frequency of a locus phenotype. Subscripts denote relationship to a genotype (ij) or to a phenotypic value (x). Note that $\bar{s}_x$, the weighted mean selection coefficient on phenotypic value, is the cost of selection; there is no meaningful "$\bar{s}_{ij}$" in the same sense—for one thing, all s_{ij} can be 0 (see text). Solid lines represent very simple relationships; dashed lines, less simple relationships. (Modified from Milkman, 1978a.)

tween phenotypic value and fitness is a graph of a fitness function. To avoid unnecessary complexity, both graphical and conceptual, it is easiest to plot fitness against a distribution of phenotypic values in which frequency density is uniform over the entire range. Such an even, or *rectangular,* distribution is seen when phenotypic values are ranked, and percentiles are useful units in this regard. The area included in this rectangular distribution is set at 1. Figure 3 illustrates several different fitness functions in which relative fitness is plotted against percentile rank of the phenotypic value. This method permits the rapid visualization of the cost of selection as well, which is the area above the line.

The first fitness function, a simple step function moving abruptly from 0 to 1, characterizes truncation selection, a process in which individuals below a threshold phenotypic value do not reproduce. Those at or above the threshold all reproduce at a uniform rate. While truncation selection is used by breeders, it is not representative of natural selection, where (in most cases) reproductive rate is an important variable.

Other fitness functions illustrated in Figure 3 are a nondecreasing one (2), which would result in directional selection; an intermediate optimum function (3), which would stabilize the phenotype; and a bimodal function (4), whose consequences would depend on the frequency distribution of phenotypes and on their genetic basis. Naturally, since any distribution can be divided into classes, these frequency functions can all be approximately represented by a table of values.

108

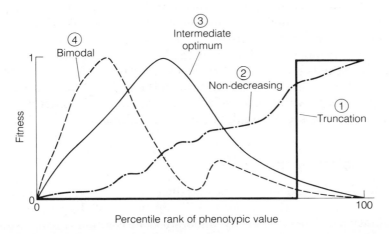

FIGURE 3. Four fitness functions (see text).

If we now graph phenotypic value directly, rather than by rank, we must consider two functions at once: the fitness function and the frequency function. In Figure 4, one fitness function (heavy line) is plotted, and it is intended to apply to all four frequency distributions. These four frequency distributions will be associated with different immediate outcomes. We can consider both the cost of selection and the direction of selection, in this intermediate optimum case. For each frequency plot, given the fitness shown, both cost and direction are

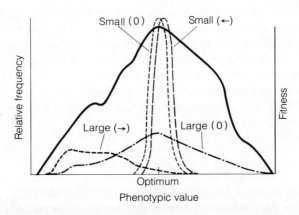

FIGURE 4. Given a particular fitness function with an intermediate optimum, four frequency distributions are shown, together with the respective costs and directions of selection that result.

109

noted. For example, the very narrow distribution centering on the optimum value will have a very low cost, since most individuals will have a fitness close to 1. Also, since the distribution is both narrow and symmetrical about a reasonably symmetrical peak of the fitness function, it is clear that directional change is near 0. On the other hand, the frequency distribution at the lower left will entail great cost (most individuals have a very low fitness relative to those in the upper range) and considerable selection toward the right. In each case $\bar{x}_s$, the mean phenotypic value after selection, is $\int x w_x f_x dx / \bar{w}_x$, or, if we group the data into classes of equal frequency, $\Sigma \bar{x}_c \bar{w}_c / n\bar{w}$.

Genetic selection models

Genetic selection models may be classified into single-locus and multi-locus, and into directional and stabilizing types. Of the four resulting combinations, three are easily understood in common terms. Single-locus directional models are of course featured in elementary texts, and they are generally followed by single-locus stabilizing models, involving heterosis (often illustrated by Hb^A and Hb^S) or frequency-dependent selection. Multilocus directional models with strict additivity are, once again, understood in the same terms, through the use of the relationship $s \approx ig$. (Interactions can introduce complexities, but these are not fundamental.) For all three models, the parameters are the same; they are used in the same way, and the dynamics and end points can be predicted. On the other hand, satisfactory analysis of multilocus stabilizing selection has not been available until recently. A satisfactory analysis in this case should be straightforward, general, accurate, and related directly to the three other categories of genetic selection models. Consider initially a model involving 100 diploid loci, each with two alleles at frequencies greater than 0.05. At each locus the allele designated with an upper-case (capital) letter makes a unit contribution to the phenotypic value under consideration, while the "lower-case" allele makes a zero contribution. The contributions are additive. One reason for the lack of a general accurate treatment of such intermediate optimum models is related to the third unification problem, namely the recognition of the collaborative influences of selection and sampling.

Selection and sampling

It is certainly obvious in recent literature and texts (e.g., Crow and Kimura, 1970) that selection and sampling must be considered simultaneously. Specifically, it is frequently pointed out that in diploid populations, selection is effective only when s exceeds $1/4N_e$, or, more

110

precisely, when Δq due to selection exceeds $1/4N_e$. N_e, the effective population size, is the actual population size diminished by excessive variance in reproductive rate, the binomial variance being taken as standard. The effect of random genetic drift on allele frequency is inversely proportional to N_e, so we are comparing Δq due to selection with Δq due to drift. Since this relationship is now widely known, our unification problem in this area clearly does not lie in the present but in the past. For before the great neutralist/selectionist controversy, most problems in population genetic theory were treated in terms of infinite populations only. Thus Robertson's skillful demonstration (1956) that intermediate optima drive dimorphic genes to monomorphism was accepted without reservation long after finite populations and random genetic drift made their forceful entry to center stage. It has now been reconsidered, as we shall see shortly.

Now that the two sets of selection-related parameters have been joined together and the collaboration of selection and sampling has been emphasized, we may proceed with the analysis of the multilocus intermediate-optimum model. Our task is twofold: to express it in terms common to the other three categories of genetic selection models, and to reconcile the coexistence of selection and genetic variation.

A major difference between single-locus and multilocus models lies in the relationship between the phenotype (implicit or otherwise) and the locus genotypes. Specifically, the obstacle to unification has been the problem of calculating fitnesses for single-locus genotypes in multilocus models. In single-locus models, the fitnesses are either assigned arbitrarily or estimated on the basis of some assumed phenotypic contribution with a consequent fitness. In multiple-locus models, the calculation of fitnesses for single-locus genotypes was somewhat obscure. Proof of the generality of the relationship $s \approx ig$ for directional selection brought that category into the fold, and the extension of the proof to the case of the intermediate optimum has now done the same for multilocus stabilizing selection. Thus, as we have seen, it is possible to calculate i as $(\bar{x}_s - \bar{x})$, where $\bar{x}_s$ is $\int xw_x f_x dx/\overline{w}_x$; to determine (or assign) g; and thus to obtain s. Recall that i and g must be in σ units, in which all further phenotypic values will be expressed.

The direct assignment of fitness values to quantitative phenotypes and the subsequent derivation of fitness values associated with locus-genotypes is of course consonant with the operation of natural (and, for that matter, artificial) selection. After all, selection operates on phenotypes directly and on genotypes only indirectly. Thus the relationship $s \approx ig$ makes it unnecessary to worry about the dependence of a locus genotype's worth on the remainder of the genotype. It is true

111

for an intermediate optimum that if all "upper-case" alleles favor large size, then the AA genotype is beneficial on a largely "lower-case" background and detrimental if the rest of the genotype is largely "upper-case." Nevertheless, the overall fitness of AA is not indeterminate, or even intractable. It is in fact easy to calculate, as we have seen.

COEXISTENCE OF SELECTION AND GENETIC VARIATION

Selection in heterogeneous environments

Before proceeding, I should point out that there is a widely accepted explanation for the existence of genetic variation in species, and it may well account for most of it. Most species ranges appear to be heterogeneous in important ways, so that selection in some sectors would oppose selection in others. Levene (1953) defined conditions under which selection in a heterogeneous environment, coupled with panmixia, would stabilize genic polymorphisms, and a considerable and rigorous descendant set of "Levene models" has extended these conditions greatly (Gliddon and Strobeck, 1975; Haldane and Jayakar, 1963; Hedrick et al., 1976; Prout, 1968; and see Jones and Probert, 1980). Yet there are two reasons to search for an even more basic reconciliation of selection and genetic variation. First, in the most uniform environments known, deep-sea trenches, genetic variation is as abundant as elsewhere, by electrophoretic criteria. Second, and more importantly, if there is a more fundamental explanation in the very nature of genetics and populations, then we should know about it.

A more fundamental explanation

We come now to the first of the three steps toward the reconciliation of selection with persistent genetic variation in the case of an intermediate optimum. Phenotypic selection first drives the population to a distribution of phenotypic values such that $i = \int x w_x f_x dx / \overline{w} - \bar{x} = 0$. When $i = 0$, then $s_{ij} = 0$ for all applicable genotypes. Thus genotypes $AA, Aa,$ and aa will all leave the same average number of progeny—they will be mutually neutral.

A qualification must be raised at once, however. In multilocus models, it is not sufficient to measure fitness to the formation of the next generation's zygotes. The variance of the progeny sets must be considered: In fact, this determines the quality, as it were, of each progeny set, as opposed to the quantity. Crow and Kimura (1970) explain it lucidly (p. 293). For example, if alleles A and B contribute one point each, and a and b contribute nothing, $AAbb$ and $AaBb$ look

112

equivalent at first glance. But $AAbb$ produces Ab gametes only, leading to progeny ranging from 1 to 3 points if mating is random (and ab and AB gametes are available), while $AaBb$'s progeny range from 0 to 4. If the optimum is 2, this difference would appear to drive out heterozygosity; this should be true for any optimum, with the potential exception of one last locus (consider an optimum of 1 in the present example). Indeed, Robertson (1956) showed that the more frequent of two alleles at each locus would be fixed in such a situation, and he showed, for an infinite population, that

$$\Delta q = \frac{a^2 pq(q - p)}{8\sigma^2}, \text{ or, in our terms, } \frac{g^2 pq(q - p)}{8}$$

The question now arises, in the modern context of random genetic drift, as to whether g is ever small enough to make this Δq much smaller than that due to sampling. The second step of this argument, then, is a reminder that in this case, as in all cases involving finite populations, the question is not whether selection is operating, but whether it is strong enough to overcome random genetic drift. If g is small enough, then the variances of progeny sets do not vary (according to the distribution of a given number of "upper-case" alleles in a parent) sufficiently to overcome random genetic drift. Thus, all the locus genotypes would be neutral.

Up to this point, we have dealt exclusively with mathematical models, and the results are unambiguous. The third step of this argument is more biological, and it favors the conclusion that g is indeed quite small in the final analysis.

To begin with, the calculation of g in the present model is simple. The total phenotypic variance, V_p, is 1 (in σ units), and we may choose to ascribe half of it to genetic causes. V_G, therefore, is 0.5, and in this additive model, all the genetic variance is additive genetic variance, V_A so $V_A = 0.5$. The variance contributed by each equally contributing locus, V_L, is V_G/L, where L is the number of loci. If there are 100 such loci, $V_L = 0.005$. Also, $V_L = pqg^2$, and we are interested in the value of g^2. If $p = 0.8$, then $g^2 = 0.005/(0.8 \times 0.2) = 0.0313$. Using Robertson's equation, Δq is then 0.0004, which would overcome random genetic drift in populations whose effective size is about 700 or larger. Thus the drive towards homozygosity would still be likely to apply in this case. Then is there a reason to believe that g^2 should actually be much smaller? There is.

An organism has only one fitness. It has many phenotypic characters, since a phenotypic character is any observable attribute of an organism. The question is whether the total phenotype is properly

113

partitioned into a number of components with all-or-none effects on survival or fertility; or whether, no matter how it is sliced, the total phenotype consists mostly of components with cumulative effects on fitness. Clearly an all-or-none susceptibility to a fatal disease determines the alternatives between death and no damage. Yet, by and large, the various phenotypic elements' effects are integrated. The same argument can be made concerning the fitness of a class of individuals, such as those carrying the AA genotype.

Further, coadaptation (big teeth, big jaws) is an obvious feature of the overall phenotype, so that the individual phenotypic values often have an adaptive significance conditional upon other phenotypic values. This all leads to the conclusion that our model should have perhaps 2,000 loci acting collaboratively, not 50 or 100. When this is the case, other values remaining as they were, $g^2 = 0.001563$ and $\Delta q = 2 \times 10^{-5}$. Note, too, that in the collaboration of many small components, pure additivity is a very good approximation.

If this biological argument is accepted, the conclusion is clear: the individual loci behave neutrally. That is, the frequencies of the alleles are influenced only by random genetic drift, for all practical purposes. Thus we arrive at the simultaneous operation of random genetic drift on allele frequencies and selection on phenotypes, when the population's mean phenotypic value, $\bar{x}$, is the same as the selected mean phenotypic value, $\bar{x}_s$. Two points should also be made here. First, the equality of $\bar{x}$ and $\bar{x}_s$ does not imply that $\bar{x}$ is also the optimum phenotypic value, since the fitness function could be asymmetric. Second, and probably needless to say, this characterization does not apply to alleles with properties different from those assigned in the model— lethal alleles, to take an extreme example.

Two senses of neutrality

It is now important to distinguish two senses in which neutrality is frequently used. The present sense, selective neutrality, refers to alleles whose frequencies relative to one another change only by chance. Another sense, phenotypic neutrality, describes two or more alleles whose substitution, one for another, would have no phenotypic impact ($g = 0$). Phenotypic neutrality results in selective neutrality; the converse is not true. Clearly, the alleles in our model are by definition not phenotypically neutral, but they are selectively neutral. Moreover, it should be clear that although the alleles are selectively neutral, the various phenotypic values are not: they have individual selective significance even when the aggregate result is the absence of change in mean phenotypic value ($i = 0$, which is to say $\bar{x}_s = \bar{x}$). It may be useful here to recall Templeton's (Chapter 2) emphasis on the deme in selection. In the present case, the deme (population) is being selected for phenotypic uniformity, and this selection is opposed by the independent segregation of alleles, which J. F. Crow calls "Mendelian noise." Though selection promotes phenotypic uniformity, there is no selection on allele frequencies when $i = 0$.

As the various allele frequencies drift randomly, some might approach 0 or 1. This, of course, is one of the possible consequences of random genetic drift. If, in a rare combination of events, enough "upper-case" allele frequencies were to rise so that $\bar{x}$ were now substantially in excess of the optimum, selection would drive it back. But the errant alleles would not be singled out for correction, of course. *All* upper-case alleles would tend to be reduced in frequency.

This scenario brings us to the current uncertainty as to the role of selection in relation to genetic polymorphism. Historical origins of the controversy aside, there remain conflicting reports on the subject today. Fortunately, this is one conflict in which both sides may be right, and the basis of its resolution has been suggested now and then over the past dozen years.

Conflicting evidence

Here are the opposing categories of evidence. Favoring neutrality is the almost universal failure of a vast array of efforts to find an adaptive basis for allozyme polymorphisms. Also favoring neutrality is the lack of suggestive distributions of allele frequencies among polymorphic loci, among populations, among environments, and over time. I should point out that there are many additional lines of reasoning that are persuasive to some, but I believe that the enduring evidence on this issue will be of a general and empirical nature.

Opposing neutrality, and thus favoring selection, is the empirical test of the steady-state equation that is the logical conclusion of the neutrality principle:

$$n_e - 1 = 4N_e u_n$$

where n_e is the effective number of alleles; N_e is effective population size (defined earlier); and u_n is the aggregate rate, at a locus, of mutation from one allele to another adaptively equivalent allele. The effective number of alleles, n_e, is the number of alleles, all of equal frequency, which would result in a given probability of picking two similar alleles successively at random. For example, if the chances of picking two similar alleles in a row were 0.25, n_e would be 4, since four alleles, each with a frequency of 0.25, would give the desired probability ($4 \times 0.25^2 = 0.25$). In addition, a particular value of n_e

115

represents other polymorphisms leading to the given probability. For example, six alleles with respective frequencies of 0.40, 0.20, 0.15, 0.15, 0.05, and 0.05 would also give the probability 0.25, and so the *effective* number would again be 4, though the *actual* number, called n_a, is 6. n_e is calculated as the reciprocal of the sum of the squared allele frequencies, $1/\Sigma p_i^2$.

The base value for n_e is 1, not 0, because there must be at least one allele at every locus. Thus the proper measure of variation is $n_e - 1$, not n_e, since there is no variation when $n_e = 1$. The equation is usually written:

$$n_e = 4N_e u_n + 1$$

which corresponds directly to its derivation, but it has clearer meaning when written:

$$n_e - 1 = 4N_e u_n$$

Since this is a steady-state equation, it applies only when a steady state has been reached, and this requires an average of $4N_e$ generations. Specifically, the values of N_e and u_n must be constant for the past $4N_e$ generations (or above a minimum level, if a minimum estimate of n_e is to be made). There is a less obvious requirement also: the alleles under consideration must remain neutral over this period of time. It is not sufficient that the *number* of neutral alternatives be constant. That would permit alleles to arise by mutation, become frequent by drift, and then be wiped out or fixed by selection, leaving their newly neutral "replacements" all at infinitesimal frequencies. So the same alleles must be neutral over the entire period needed to reach the steady state.

This equation was tested by me (1973) in *E. coli,* one of the few organisms for which there is strong evidence that the proper number of generations have elapsed. Because *E. coli* is haploid, the equation is $n_e - 1 = 2N_e u_n$. In any event, the observed and predicted values of $n_e - 1$ differ by 2 to 3 orders of magnitude. Selander and Levin (1980) demonstrated a similar disparity with a larger number of loci but attribute it to an overestimate of N_e. They consider that periodic selection keeps N_e lower, and they offer some concrete evidence for this contention. Furthermore, Ohta (1974) and Kimura (1979) have added purifying selection to the neutral model in ways that minimize the dependence of $n_e - 1$ on N_e. As Selander points out (1976, p. 42) this still does not conform to the selectionist basis of polymorphism, which is the favoring of each polymorph under a particular set of conditions. Purifying selection only reduces allele frequencies. Ohta and Kimura say that one allele (or a set of alleles) is best, and many others are almost as good. Selectionists say that all alleles present at frequencies above 0.01, for example, are favored by selection under some circumstances.

116

Resolution

In any event, and in spite of the existence of pertinent counterevidence and counterarguments, I believe that the current balance of evidence is inconsistent with *eternal* neutrality. This leaves us with one possible conclusion that is consistent with all major lines of evidence: most alleles at frequencies above 0.01 are neutral now, but each has been favored at some time in its history.

This state of affairs bears an amusing similarity to the simple-minded, and of course improper, explanation of genetic variation—that it's there so that the species can cope with changing conditions in the future. Indeed, the genetic variation, neutral at present, is capable of responding to selection, should the optimum alter.

A POSTSCRIPT

I would like now to turn to the evidence for the role of selection in evolution on the grand scale. In one way or another, the theory of natural selection is an extrapolation of microevolutionary selection events, artifical or otherwise. This extrapolation consists either of enlarging the steps (macroevolution) or prolonging the path (many little changes add up to a big change). Obviously, speciation is an important stage in the process, and it also seems to represent a stage which our understanding (not to mention experimental capability) has not yet reached. We have probably not witnessed the origin of new species (reproductive isolates artificially maintained do not count), as Bush points out in Chapter 7, and we have certainly never created real species experimentally.

In fact, there is a nearer stage that we have not reached experi-mentally, and it is worth a few words of description. When we select for phenotypic change, we ordinarily get a response (provided the population contains relevant genetic variation, and most do). When we relax (abandon) selection, we get either a reversal of the response, or the absence of directional change. This general pattern is inter-preted as natural selection favoring a return to the original optimum, or point of greatest stability; a failure to return is attributed to the elimination of relevant genetic variation in the course of selection. The point is that the optimum, the center of stability, has not been affected materially by selection, and yet this is clearly a critical step toward speciation. To my knowledge, we have never achieved it ex-perimentally.

To do so would require substantial alteration of a (quantitative) phenotype and its stabilization, perhaps by a lengthy existence after

117

fixation of the alleles selected. During subsequent generations, selection should favor accommodation of the fixed phenotype. Eventually, back selection could be instituted if relevant variation were restored, perhaps by crossing two or more lines with different fixed genotypes. Now, after a return halfway to the original value, what would happen if selection were relaxed? A change in the direction of the more recent value would indicate the change in the point of maximum stability, the genetic reorganization necessary for speciation.

CHAPTER 7

WHAT DO WE REALLY KNOW ABOUT SPECIATION?

Guy L. Bush

The diversity of life on this planet is primarily the direct outcome of a process called speciation, the splitting of one species into two, each of which represents a closed genetic system free to follow its own evolutionary destiny. Although the importance of speciation is clear and convincing, the processes involved are, for the most part, unknown. Evolutionary biologists have not even been able to agree on a suitable definition of a species, a factor that in some respects has hindered our progress in unraveling the details of how speciation occurs in nature. Invoking one definition or another tends to determine how one views the process and the importance of various factors in speciation.

There is no question that the end product of speciation is effective reproductive isolation between natural populations. But reproductive isolation is after all the end product, not the cause, of speciation. It is really the process of speciation that concerns us, and this is what we know least about. It is no wonder the subject has generated so much speculation and controversy over the years.

Our lack of understanding possibly stems from the fact that the origin of each new animal or plant species represents a unique series of evolutionary events. The processes involved in speciation are never exactly duplicated even within a lineage. Each population with its own special gene pool experiences a constellation of different patterns of mutation, gene flow, selection, and drift. It is this uniqueness, this never-to-be-repeated-again quality that makes it difficult if not impossible to formulate realistic all-encompassing models of speciation, or to devise suitable experiments to test our theories. Furthermore,

119

speciation is usually a rare event, seldom if ever observed from start to finish. Our current concepts of speciation are therefore primarily based on *post hoc* reconstructions of past events, or derived from theoretical population genetic models usually based on classical Mendelian genetics, with all the inherent weaknesses and speculative nature of these approaches. The *post hoc* approach is, at best, subjective and it is thus not surprising that recent advances in molecular biology call into question certain widely held conclusions of the naturalists and population geneticists (Crick, 1979).

In the past, there has been a procrustean tendency to cast all modes of speciation in geographic terms (i.e., allopatric, parapatric, sympatric, etc.), a reflection no doubt of our inadequate knowledge of what actually goes on during speciation. There is no question that geographic patterns of distribution affect the mode and rate of speciation. This view has long been the theme of the Neodarwinian school (cf. Mayr, 1963) and recently taken up by the vicariance biogeographers and cladists (Eldredge and Craycraft, 1980) as the ultimate solution to the speciation problem. But knowing the pattern tells us little about the actual processes involved. Populations with very different gene pools and isolated for millions of years may show no signs of speciation when tested, while sympatric forms exhibiting little genetic or phenotypic differentiation may be completely reproductively isolated.

Because it is the process and not the product of speciation that remains an enigma, I would like to focus attention on several outstanding problems that must be resolved if we are to understand the way new species arise in nature.

THE GENETICS OF SPECIATION

One has only to peruse the literature to realize that although much has been written, little concrete information is actually available on the genetics of speciation. For instance, a genetic cornerstone of current speciation theory is "coadaptation." Dobzhansky (1951), Mayr (1954), Carson (1959), and many others have championed the view that natural selection inevitably favors combinations of alleles at nearly all loci that must harmoniously interact. They claim that this coadaptation of the gene pool results in strong genetic cohesion that resists evolutionary change. Furthermore, because of the coadapted nature of the gene pool, any change in gene frequency or loci will affect all other genes and thus alter their selective values. Therefore, speciation, they argue, must necessarily be slow in order to allow time for the readjustment of the genotype at a new peak, a process that must inevitably involve the accumulation of many small changes in gene frequency (micromutations) at many loci. Rapid speciation is thus precluded, except under very special conditions involving small founder populations and "genetic revolutions." The concept that the

120

gene pool is highly coadapted and exerts a strong cohesive force against genetic change has also been invoked to reject the idea that mutations in a few key loci can result in a rapid adaptive shift and phenotypic change.

Unfortunately, the hard data on which the concept of coadaptation is based are not impressive. Specific cases where the number, rate, kind, and mode of action of genetic loci involved in speciation have been established are woefully lacking, and I am unaware of any unequivocal cases demonstrating that genetic revolutions have been directly associated with speciation. In fact, in cases where it has been expected, such as the Hawaiian *Drosophila,* which are likely candidates for speciation by the founder effect, it has not been found (Carson and Kaneshiro, 1976; Templeton, 1982).

In almost all cases, the observed facts are controversial and, upon close scrutiny, clearly open to conflicting interpretations (Nei, 1981). As emphasized by Hedrick et al. (1978) in a detailed review of the problem, coadaptation has often been invoked in numerous speculative writings to explain the origin and retention of multilocus systems, but rarely has it been tested experimentally or even demonstrated in nature. In fact, selection at individual loci rather than coadaptation between loci can satisfactorily explain most, if not all, cases of gametic disequilibrium (Hedrick et al., 1978). Even studies claiming genetic evidence of coadaptation, such as those showing a nonrandom association between electromorphs and inversion chromosomes (Prakash and Lewontin, 1968), can be explained by initial linkage disequilibria generated when the inversion is formed or by random genetic drift. The results of such studies can be accommodated within the neutral mutation hypothesis (Nei and Li, 1980).

The amount of genetic variation in most higher organisms is now known to be astonishingly high, and this further weakens the case for extensive, all inclusive coadaptation. As pointed out by Kimura (1974), the paired DNA blueprints of each diploid organism are recombined more or less at random each generation. Human haploid genomes differ from one another on the average of at least a half-million nucleotide pairs. Some of this variation may be present in highly reiterated sequences, but much appears to exist in genes coding for proteins. Is it reasonable to assume that natural selection can mold this variability into a single set of finely tuned coadapted gene complexes throughout the entire gene pool of the species? Does it really have to? Is the variation we uncover in our electrophoretic surveys or assessments of nucleotide sequences of structural genes of much real importance in adaptive evolution and speciation?

For instance, what is the adaptive function of a paracentric inver-

121

sion occurring at high frequency in a natural *Drosophila* population? Is it, as claimed by advocates of coadaptation, a chromosome rearrangement that protects coadapted gene complexes within the inverted area from recombination? Or might we interpret past studies showing overdominance of inversion heterozygotes as examples of simple regulatory polymorphisms in which the inversion serves to alter some basic pattern of gene regulation, thus affecting the expression of many linked, or possibly even unlinked, genes simultaneously? One only has to review the phenotypic effects of chromosome rearrangements in such diverse organisms as *Drosophila* (Lindsley and Grell, 1968) and man (Borgaonkor, 1980) to realize that chromosomal mutations can have a profound effect on the phenotype. In some well-studied cases, such as sex determination by the HY antigen, the phenotype may be radically altered without a single change in the structural gene (Koo et al., 1977).

This is not to say that certain patterns of gene association and linkage may not evolve, such as super genes like the *pin* and *thrum* phenotypes of *Primula,* the color pattern genes of the grouse locust, mimicry in certain butterflies, and other examples treated in detail by Ford (1975). Clearly, with the discovery of transposable elements, there are sufficient molecular mechanisms for reorganizing the genome on a micro or macro scale through chromosome mutations, bringing genes together in functional relationship. Such super genes may also be regulated by a single control point. Although multilocus control may exist with or without tight linkage (Brown, 1981), it does not necessarily require coadaptation of different allelic variants of gene products across the entire genome in the classical sense, but it may merely involve precisely timed patterns of gene expression.

If coadaptation plays an important role in adaptive evolution and speciation, it may well be at the level of the various control points in gene regulation rather than between genes coding for proteins. At least this might possibly reduce the level of coadaptive complexity to a manageable level, because a single regulatory locus may control a large number of structural loci. However, even at this level, it is quite possible that locus-by-locus selection may play a predominant role. There simply is no convincing evidence either way. This interpretation is just as consistent with the results of past studies as the conventional one of genic coadaptation. Unfortunately, the concept of genic coadaptation has now become so ingrained in our evolutionary lore that we tend to accept it as a proven fact. This in turn has placed several constraints on the way we have interpreted evidence generated from our evolutionary studies on the genetics and population dynamics of natural populations. In particular, it has had a profound, and I feel inhibitory, effect on how we view speciation.

122

IMPORTANCE OF GENE REGULATION

The problem of coadaptation and its supposed cohesive effects leads directly to, and is really inseparable from, another problem that has been given considerable attention lately, namely macroevolution. Goldschmidt (1940), at odds with the conventional interpretations of the speciation process, argued that new species could arise in a very short time, not as a result of the accumulation of many small adaptive "micromutations" in coadapted gene complexes but by one or more "systemic" or "macromutations" that produced an evolutionary "hopeful monster." This view was not a popular one among his contemporaries who dismissed speciation by macroevolution as an adaptive impossibility. After all, what chance would one of Goldschmidt's hopeful monsters have of meeting and mating with another hopeful monster with an identical macromutation and starting a new species on its way?

The unfortunate coining of the term "hopeful monster" to describe the product of his macromutations may have clouded the picture somewhat, as it conjures up visions of some grotesque *Missbildung* that would have almost no chance of surviving in nature, let alone finding a mate with like traits. But a rereading of Goldschmidt today leaves a quite different impression, particularly in the light of recent discoveries in molecular genetics.

Goldschmidt's systemic mutations represented special alteration of the genome that changed the primary pattern of what he termed *reaction systems* controlling development. He was rather specific as to the kinds of mutations that were likely candidates for macromutations. These involve chromosome rearrangements which result in chromosome repatterning of the genetic material. He also postulated that one or more macromutations could survive only under the right circumstances, such as the absence of strong selection pressure against the heterozygote and inbreeding. This view is consistent with the current theory on the fixation of chromosome rearrangements (Wright, 1941; Lande, 1979). Such mutations are what we would now recognize as regulatory mutations that alter the timing and pattern of expression of several genes simultaneously, producing a phenotypic change.

For various reasons, Goldschmidt abandoned the strictly corpuscular (i.e., beads on a string) concept of the gene that was becoming widely accepted in the 1930s because he was convinced from his developmental genetic studies and other evidence that the genetic material consisted of chromosomal proteins that were ordered into a single long molecule on each chromosome with regions of specific

123

function along its length. Furthermore, he took his model one step further and erroneously concluded that all mutations arose by chromosome rearrangements. Point mutations, in his view, did not exist, a view that was clearly wrong. I should add, however, that his critics were also wrong in attributing all adaptive evolutionary change to point mutations in the genetic material. Chromosome mutations were and still are viewed by many today, despite much contradictory evidence, as playing at most a minor role in the adaptive process and speciation.

You might say that Goldschmidt, like his contemporaries, was "half-right." His view that chromosome rearrangements play a special role in evolution and speciation is now increasingly supported by evidence emerging from the discoveries on eukaryotic chromosome and on the complexity of gene structure and organization by molecular biologists (Bush, 1981; Flavell, 1981). Indeed, there is some indication that genes may even "talk" to each other by way of introns in the process of coordinating expression of multigene systems (Lewin, 1981), and that the intervening sequence of one gene may serve as the coding region for another with which it interacts (Borst and Grivell, 1981). Certain genes, such as those responsible for the immune system in vertebrates, even rearrange themselves in a specific manner within an organism in order to augment immune specificity (Davis et al., 1980b). Thus, a single nucleotide sequence is used in two or more ways. Furthermore, the discovery of overlapping genes in certain viruses opens the possibility that similar patterns of gene expression may exist in higher organisms. Recent studies by Hagenbüchle and Schibler on mouse α-amylase genes suggests that the same DNA sequence produces mRNAs which differ between liver and salivary gland only in that one is extended by about 100 nucleotides (Minty and Newmark, 1980). All this suggests that genetic organization from the standpoint of gene regulation, development and phenotypic expression may resemble Goldschmidt's holistic view of the genetic material more than his contemporaries would care to admit.

The picture that is emerging is that nature has devised many alternative ways of regulating gene expression in higher organisms (Brown, 1981). These include the direct alteration of genes through diminution, amplification, and rearrangement, as well as the modulation of gene expression by transcriptional, posttranscriptional and translational processes at several *control points* in the nucleus and cytoplasm. In some cases, many genes may be controlled simultaneously or in sequence.

It is also increasingly clear that many eukaryotic genes are arranged in multigene families. Some simply represent reiterated copies in tandem along a chromosome. Others may have several related genes in a repeating unit, while more complex multigene families may link genes under developmental control. In the final analysis, the number of control points will quite likely be far fewer than the number of structural genes producing a product, but the picture is still unclear. Finally, evidence has been presented suggesting the inheritance of certain types of induced phenotypes (see Chapter 11). These Lamarckian-type phenomena have been observed in diverse physiological systems, and, if substantiated, they will have a profound effect on how we view evolution and speciation. The picture now emerging at the molecular level is thus quite different from our perception of gene action and control we accepted even five years ago.

Where does this leave us with respect to current speciation theory? If coadaptation and its postulated genetic cohesiveness is not the important limiting force on adaptive change we thought it was, what kind of genetic changes might be tolerated? Can Goldschmidt's macromutations result in rapid evolutionary change and speciation after all? I think the answer is a *qualified* yes.

My qualifications are directed more toward the interpretation of what constitutes a macromutation than the concept itself. Over the years, macromutations have been equated with gross morphological change. Although hopeful-monster-type mutations clearly occur and are well documented not only by Goldschmidt himself (1940) but also in medical literature, they represent only the most conspicuous expression of macromutations. In my view, a macromutation is any mutation that results in a major adaptive shift, a change in the way of life that opens up a new adaptive zone.

Such mutations may or may not be expressed as a visible morphological change. As an example, a chromosome rearrangement or a point mutation affecting a key control point arising in a small inbreeding population that results simultaneously in the derepression of a digestive enzyme and reduced fitness of heterokaryotypes may be sufficient to initiate a speciation event. A critical digestive enzyme in short supply may, when abundant, permit a population fixed for the new rearrangement to exploit food resources previously off limits, thus conferring enhanced fitness. Such a mutation would reduce competition with the parent species as it represents a significant adaptive shift, further enhancing the prospects for the development of reproductive isolation (Rosenzweig, 1978; Bush, 1981).

The fixation and survival of a macromutation will require very special conditions indeed, such as small effective breeding size, inbreeding, and drift (Wright, 1941; Lande, 1979). Such conditions are not often precisely met in nature, but as my colleagues and I have pointed out elsewhere, some groups of organisms, because of their

125

peculiar biologies and population structures, are more prone to rapid evolution through the fixation of such mutations than others (Bush, 1975; Wilson et al., 1975; Bush et al., 1977; Bush, 1981).

The argument is frequently voiced that although macromutations are often observed in laboratory populations and occasionally in nature, none appears adaptive. In not one case have such macromutations ever resulted in an adaptive shift or speciation. But is this evidence that evolution and speciation cannot occur by macromutation and macroevolution? The argument, I think, is spurious. A consideration of speciation rates will illustrate this point.

We have estimated that horses and some primates are the most rapidly speciating groups among the vertebrates (Bush et al., 1977). This "rapid" rate, however, represents only one speciation event every 50 to 200 thousand years. The average rate for most other vertebrates is one speciation event every 2 to 3 million years. Even in the unlikely event that in vertebrates (a phylum with some 40,000 extant species) all speciation events occurred by way of macromutations, the chance of recognizing and observing even one in the lifetime of an evolutionary biologist is infinitesimally small. Not only would some involve no morphologically observable change, particularly in the early stages, but the chance of being at the right place at the right time and with the appropriate history of the population to understand and interpret the events when they happen is, to say the least, a matter of considerable luck or clairvoyance. Insects, which number over a million species and probably speciate at higher rates, should offer a better chance to observe speciation, but, at least in many cases, their inconspicuous nature precludes easy study.

To document a speciation event, it is necessary to observe a large number of animal and plant species in depth, such as has been done with *Drosophila,* for a number of years. We are, in fact, ignorant of the biology of most organisms except for superficial attributes. This is particularly true of parasites, which probably outnumber all other plants and animals together.

SPECIATION AND ADAPTATION

Another sacred cow of Neodarwinism and a by-product of the cohesive-gene-pool school is the idea that speciation is the by-product of many fortuitous genetic changes, with small effects incorporated into the gene pool of a population as a result of selection and drift during periods of geographic isolation. Only if the right combination of genes diverge will isolated populations remain reproductively isolated if and when contact is reestablished, although reproductive isolation might be strengthened later by selection. The evolution of reproductive isolation and speciation is thus not directly part of the adaptive process that results in divergence. Unfortunately, we know rather little about

126

what kind of genetic changes are involved in the development of reproductive isolation. Do these micromutations occur at control points, in structural genes, or in both? How many are necessary for speciation, and how are they established and spread? What role, if any, do the various classes of repeated DNA sequences as well as introns and transposable elements play in the process? What is the relative importance of selection and stochastic forces in determining the pattern of divergence? And, a question already raised, is genomic coadaptation really a factor of overriding importance? These are only a few unresolved problems that come to mind, none of which has been satisfactorily answered.

There is no question that speciation does occur as a result of geographic isolation. But how important has it really been? We only have score cards inferred from rather circumstantial evidence, which in turn has been derived from a few animals and plants that speciated sometime in the past. It is for this reason and others alluded to earlier that I remain skeptical of the universal application of allopatric speciation to all sexually reproducing eukaryotic organisms.

There is an alternative hypothesis that has been proposed in one form or another over the years, in which reproductive isolation is directly selected for in response to strong divergent selective pressures exerted on two or more populations specializing in different habitats (for reviews see Bush, 1975; Endler, 1977; White, 1978; Templeton, 1981). In these models, speciation is the direct outcome of adaptation, and divergence occurs as a product of selection for habitat preference, competition, and selection to enhance reproductive isolation. Geographic isolation in this case is not a prerequisite for the evolution of reproductive isolation which arises as part and parcel of the adaptive process during speciation. Only in the last few years have we begun to explore the possibilities of this mode which, for want of a better term, might be called *adaptive speciation*.

It is clear that these nonallopatric models of speciation suffer from the same lack of hard data on the genetics of speciation that plagues the proponents of allopatric speciation. Whether any of these models reflect natural processes has not been conclusively established. If a few mutations can result in adaptive shifts of major consequences, then there appear to be no major obstacles to the occurrence of some form of nonallopatric mode of speciation in natural populations. Much hinges on the reality of coadapted gene pools and the effect they have on what kind of mutations are acceptable in evolution, i.e., micromutation with small effects or macromutations with significant adaptive shifts.

I believe the proponents of the allopatric and nonallopatric schools

127

have their work cut out for them. It is no longer acceptable for the advocates of allopatric speciation to dismiss nonallopatric models of speciation on the grounds that they lack convincing evidence when the counter arguments they themselves use are tenuous and based on equivocal evidence and speculation. Rudyard Kipling might recognize an old familiar pattern in our attempts to explain the way new species arise in nature. But alas, the time has passed for more "Just So Stories," including a diverse group of recent vintage (none as scientific as they purport to be).

We are on the threshold of a new understanding of speciation and evolution. It is instructive, I think, that the evidence which will enable us to cross that threshold no longer comes exclusively from population genetics, field biology, and systematics but from the information emerging in abundance from the benches of the molecular biologist. As evolutionary biologists, our view of genetics has been limited because we lacked sufficient knowledge about gene structure and function. The molecular level was until recently treated like a black box where anything was possible or limited depending on the theoretical paradigm. It now appears that we will soon learn a great deal about the molecular attributes inside that black box and their role in development, adaptation, and speciation.

Obviously, the future holds many surprises; evidence has come from unexpected areas of research in molecular genetics and will continue to do so. As to the outcome, I suspect that macromutations and rapid nonallopatric mechanisms of speciation will prove to be far more important in many groups of organisms than previously imagined. Ultimately, all our models must be recast in a molecular context; also, we must describe gene flow, selection, drift, and competition (and their consequences) in more exact terms. Only then can we formulate testable hypotheses of speciation. Until we know more about the molecular machinery of adaptation—that is, what is and is not possible at the molecular level—our models of speciation must remain little more than speculation based on the subjective interpretation of equivocal data. I look forward with anticipation and excitement to the dramatic reshaping of our understanding of speciation which is now in progress.

CHAPTER 8

ORIGINS OF PROTEIN STRUCTURE AND FUNCTION

Bryce V. Plapp

X-ray crystallographic, chemical, and kinetic studies during the past 20 years have led to considerable knowledge about how protein structure is correlated with function. These studies have been directed toward several questions. How does the function of a protein, enzymatic catalysis, for example, emerge from a linear sequence of amino acids? How does a polypeptide chain fold up, interact with cofactors or coenzymes, and specifically accelerate the chemical reactions of substrates by perhaps 10 orders of magnitude? How, furthermore, are such attributes as stability, enzymatic specificity, and regulatory activity built into a protein? The answers are not yet complete, but I shall discuss in this chapter the principal findings and conclusions, which have implications for studies on evolution.

The functions of proteins originate in their structures, as described by the following statements. The linear amino acid sequence is often processed, posttranslationally, to become a modified chemical structure. The polypeptide chain folds spontaneously to form a unique three-dimensional structure, which may be closely related to structures with dissimilar sequences. Proteins are often aggregates of subunits that may cooperate in generating function. The native structure of a protein is flexible and usually changes conformation when subunits or cofactors bind. Many amino acid residues participate in binding substrates into the active site, but only a few act directly in catalysis. Changing an amino acid residue in the sequence can affect the structure and function of a protein in a variety of ways.

PROTEINS ARE PROCESSED

As a protein is synthesized on a ribosome, or shortly after, its structure may be chemically modified (Uy and Wold, 1977; Schulz and Schirmer, 1979). One type of processing involves the peptide backbone. The amino-terminal formyl or methionyl groups may be removed and the new amino group acetylated. This probably occurs, for example, with horse liver alcohol dehydrogenase, since it has an amino terminal N-acetylseryl sequence (Brändén et al., 1975). Furthermore, the gene for *Drosophila* alcohol dehydrogenase has been isolated and sequenced, and it codes for a methionylseryl sequence, but the protein probably begins as an N-blocked serine (Benyajati et al., 1981). The polypeptide chain may also be cleaved by endopeptidases, such as in the removal of signal peptide sequences, which determine if a protein is to be transported through a membrane and perhaps exported from the cell. Such proteolysis may expose an aminoterminal glutamine residue, which can cyclize to form a pyrrolidone carboxyl group, as in silkmoth chorion proteins (Regier and Kafatos, 1981). Several proteins, e.g., insulin and pancreatic proteases, are also synthesized as zymogen forms and converted into their active forms by proteolysis.

The amino acid side chains may also be modified by the addition of carboxyl, hydroxyl, amido, acetyl, phosphoryl, methyl, glycosyl, or halo groups. As far as I know, none of the above modifications directly affects the catalytic center of the protein, but the modifications may be involved in regulating the catalytic activity or in maintaining the protein in a functional state. For instance, γ-carboxylated glutamic acid residues in blood clotting factors bind calcium ions and hold the proteins onto a phospholipid matrix while the clotting factor is activated. Some cofactors may be attached covalently, such as heme to cysteine residues in cytochrome c or flavin adenine dinucleotide to a histidine in succinate dehydrogenase. Hydroxyamino acids may be converted to α-ketoacyl groups, which can function as catalytic groups in active sites. About 140 different amino acid derivatives are now found in proteins, greatly increasing the variety obtained with 20 amino acids (Uy and Wold, 1977). Other cofactors, such as metal ions and coenzymes, bind noncovalently.

PROTEINS SELF-ASSEMBLE

Before, during, or after a protein is processed, it can fold up spontaneously into the structure that we find for the native and catalytically active enzyme. That this process is spontaneous greatly simplifies the problem of the translation of genetic information into a functional entity, because (in general) "helper" proteins are not required for the folding process. The classic experiments showed that the denatured

130

and reduced form of pancreatic ribonuclease can fold up spontaneously in the presence of air to form the native structure. The correct pairing of the disulfide bonds indicates that the structure has folded up correctly, but the disulfide bonds themselves do not direct the folding process. The same type of experiment performed by Stellwagen and Schachman (1962) on the multimeric enzyme, aldolase, showed that even proteins with more than one subunit fold up spontaneously. More recently, it has proved possible to renature the dimeric liver alcohol dehydrogenase when the appropriate concentrations of coenzyme and zinc ions are present (Gerschitz et al., 1978).

The folding process is not irreversible; instead there is an equilibrium between the native and denatured forms which is delicately balanced, since the overall free energies of these two states differ by only some kilocalories. Thus it is easy to denature a protein with a perturbation of the solvent conditions or temperature. It is the cooperative interaction of many small forces that determines the final folded or unfolded state. These include hydrophobic forces, which arise because large nonpolar amino acids prefer to be buried inside the "wax ball" of the protein rather than to be exposed in the aqueous solvent. On the other hand, various hydrophilic amino acids, with polar or charged side chains, prefer to be on the outside of the structure in contact with the solvent. Hydrogen bonding interactions between the amino acid side chains or the peptide bonds in the backbone help specify the particular folding pattern.

The pathway of folding is now an interesting subject of investigation. The process is quite complicated, usually going through various intermediates and abortive side reactions, which slow down the process. In spite of this, the folding is kinetically very fast, so there appear to be only a small number of pathways. Since folding is fast, it is clear that not all possible pathways leading to all possible conformations are tried, and thus the native protein obtained may not be the most thermodynamically stable one. Folding is a cooperative process that probably begins with the collapse of the random coil, so that hydrophobic amino acids are buried in the interior of the structure and so that some α-helices are formed. Sites of nucleation form, and the structure anneals until the "native" stable structure is obtained.

THERE ARE COMMON FOLDING PATTERNS

Many proteins have been studied by x-ray crystallography and are found to have α-helices or β-structures as common elements of secondary structure. More interesting are the "super-secondary" and ul-

131

timately the tertiary structures of the protein, where it appears that related proteins may have related folding patterns or that even quite different proteins can have similar three-dimensional structures.

There are five structural classes (Levitt and Chothia, 1976; Schulz and Schirmer, 1979). A few proteins have almost entirely α-helical structures. The α and β subunits of hemoglobin, myoglobin, and the protein hormones, insulin and glucagon, are among these. It was quite gratifying to see that the prediction of the α-helix by Linus Pauling was correct when the first protein structures were solved by x-ray crystallography. Although few proteins contain only α-helical structures, most proteins have some α-helical structures.

The second class of proteins has almost entirely β-pleated sheet structures. These strands of β-structure may associate in parallel or antiparallel manner, or be some combination of these, where the direction of a strand, by convention, is from the amino terminal to the carboxyl terminal. An example of this class of proteins is the enzyme copper-zinc superoxide dismutase, which has two β-sheet structures that are folded on top of one another. It is interesting that a similar folding pattern is also found in each of the domains of the immunoglobulins, although they do not have copper and zinc bound to them and do not have enzymatic activity. Other proteins in this class include concanavalin and prealbumin. Portions of other enzymes also have β-sheet structures.

The third class of structure is one in which there are both α-helix and β-sheet structures, but where these are segregated along the chain. There are many proteins in this class, which includes pancreatic ribonuclease, carbonic anhydrase, lysozyme, staphylococcal nuclease, and bacteriochlorophyll.

The fourth class of structure has the α and β structures alternating in the sequence. There are many proteins in this class and a beautiful example of one such structure is the triosephosphate isomerase molecule (Figure 1). There are eight strands of parallel β-pleated sheet structure that form a barrel that is flanked by α-helices. In fact, this very regular structure, called the TIM barrel, is also found in some other enzymes, such as pyruvate kinase and glycolate oxidase from spinach. All of these enzymes have quite different catalytic activities, and it is remarkable that these enzymes should have the same general folding pattern.

Another type of alternating α-helix and β-sheet structure is that found in one domain of lactate dehydrogenase. The nucleotide binding domain of this enzyme has a β-sheet of six strands of parallel β-structure, which are connected by two pairs of α-helices and some random coil structures (Figure 2). The coenzyme is bound at the carboxyl terminal end of the β-pleated sheet structure. Very similar

132

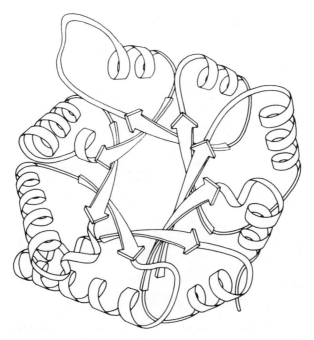

FIGURE 1. End view of triosephosphate isomerase; "TIM barrel." (Courtesy of Jane S. Richardson.)

structures are found in several dehydrogenases and kinases, proteins that bind nucleotides (Rossmann et al., 1975). These include alcohol, lactate, malate, and glyceraldehyde-3-phosphate dehydrogenases as well as phosphoglycerate kinase, phosphoglucomutase and adenylate kinase. In Figure 3, the coenzyme binding domains of three related dehydrogenases are represented in stereoscopic form. Since almost all of the amino acid residues that participate in binding the coenzyme in the dehydrogenases are *not* conserved from one structure to another, it is not clear why the nucleotide binding domains should have similar supersecondary structures. (One can imagine that coenzyme binding sites could be constructed in a variety of ways.) Probably, these proteins are related evolutionarily, but the nucleotide binding domains may have some essential physical-chemical features. In any case, these proteins, which have little or no sequence homology, have essentially the same folding pattern. The standard significance of relatedness for the six-stranded β-sheet topologies in the dehydrogenases has been

133

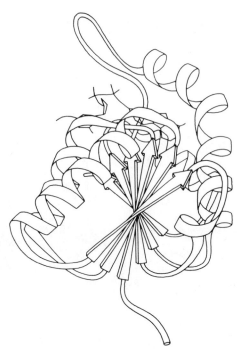

FIGURE 2. Domain 1 of lactate dehydrogenase, showing the binding of coenzyme at the carboxyl terminal end of the β-pleated sheet structure. (Courtesy of Jane S. Richardson.)

estimated to be 3500. Thus, the probability of finding this topology in two different proteins is 1/3500, and this is accepted as convincing evidence that there is an evolutionary relationship among these proteins (Schulz and Schirmer, 1979). On the other hand, an extensive study of β-sheet topologies has indicated that the number of highly favorable folding patterns is quite limited and that topological simi-

FIGURE 3. Stereo plots of the coenzyme binding domains for A, liver alcohol ▶ dehydrogenase; B, lactate dehydrogenase; and C, glyceraldehyde-3-phosphate dehydrogenase. (From Ohlsson et al., 1974.) This drawing and the stereo pairs in Figures 5, 6, and 12 may be observed with a stereoscopic viewer, available from Abrams Instrument Corp., Lansing, Michigan. Alternatively, with some effort you can allow your eyes to diverge and focus in the distance and try to fuse the images in the middle. You may find that it is helpful to put a white card between the two structures or to touch your nose to the paper between the pictures and slowly move the paper away from your eyes. If you cross your eyes, a three-dimensional image can be formed, but the structure will appear to be the mirror image of the correct stereoisomer.

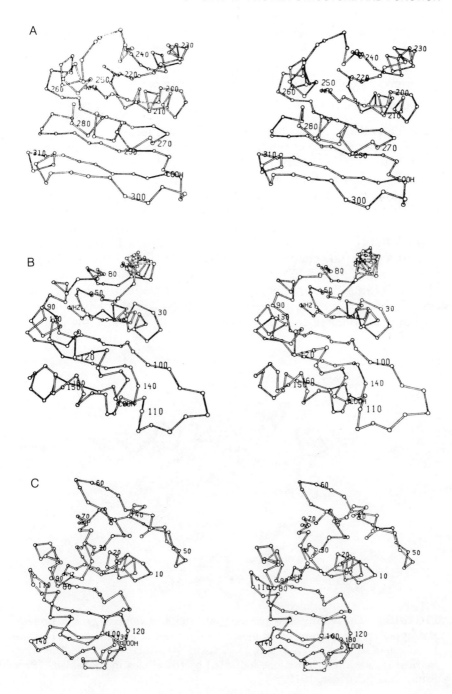

larities among other proteins can happen by chance (Richardson, 1977).

The fifth class of protein folding structures are those that contain no α-helix or β-sheet structures. Phospholipase is an example of this class.

THREE-DIMENSIONAL STRUCTURES REVEAL EVOLUTIONARY RELATIONSHIPS

In order to demonstrate further that common folding patterns can be significant indicators of relatedness among enzymes, we may consider the structure of pancreatic α-chymotrypsin. This protein has extensive β-sheet structures that are organized into two small β-barrels. Another protease, protease B from *Streptomyces griseus* has a very similar folding pattern. As shown in Figure 4, however, some portions of the chymotrypsin molecule have been changed in the protease B structure. All of the insertions and deletions are at the molecular surface, so that the amino acid residues in the inside of the structure have not been greatly altered. Even though these two proteases have very similar folding patterns, only 13 percent of the amino acid residues are identical after the two structures are geometrically aligned. It should also be noted that both proteases have the same arrangement

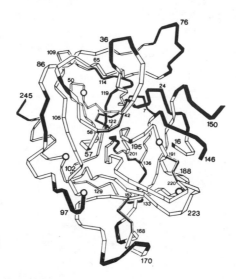

FIGURE 4. The folding of α-chymotrypsin and *Streptomyces griseus* protease B. Chymotrypsin is represented by a ribbon kinked at each C$_\alpha$-atom. Protease B is superimposed on the α-chymotrypsin structure except for deletions marked in black and insertions indicated by the circles in the ribbon. (From Schulz and Schirmer, 1979.)

136

of amino acid residues in the active site, which involve histidine residue 57 and serine residue 195.

Another, more bountiful set of proteins with similar folding patterns and distantly related sequence homologies are cytochromes c from horse, tuna, and four photosynthetic and respiring bacteria. The three-dimensional structures differ because of insertions or deletions in the sequences, but amino acid residues critical for interactions with the heme are conserved. Only after the three-dimensional structures were determined could the sequences be aligned properly (Dickerson, 1980).

The most extreme example of similarity in folding patterns in the absence of detectable sequence homology is seen in the comparison of lysozymes from bacteriophage T4 and hen egg white. Matthews et al. (1981) have concluded that the similarities in folding of the polypeptide backbone are statistically significant. By an appropriate transformation, it is possible to superimpose parts of the molecules, which happen to be near the active sites of the enzymes (Figure 5). The rest of the structures are different because of changes, insertions, or deletions in the sequences. Furthermore, oligosaccharides bound into the active sites closely superimpose, as do functional groups of amino acid residues that participate in saccharide binding (Figure 6). It appears that the catalytic mechanisms of the two enzymes are similar, even if the T4 lysozyme is about 250-fold more active than the chicken lysozyme toward *E. coli* cell walls. On the basis of this evidence, Matthews

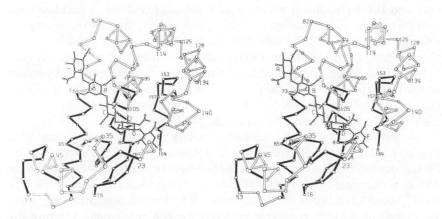

FIGURE 5. Backbone of T4 phage lysozyme with saccharide bound in subsites A, B, C, D, and E. The α-carbons equivalent to those in hen egg white lysozyme are connected by solid lines. (From Matthews et al., 1981.)

137

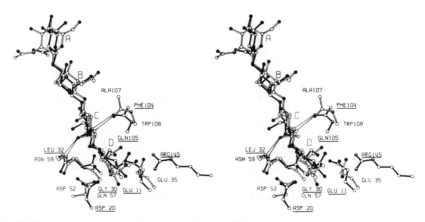

FIGURE 6. Common elements of the phage lysozyme active site superimposed on the active site for hen egg white lysozyme. Hydrogen bonds to the acetyl group on saccharide C are indicated by dotted lines. The solid bonds and the underlined names are for the phage structure. (From Matthews et al., 1981.) A more detailed view of the active-site of the phage lysozyme is available in Figure 7 of the article by Anderson et al., 1981.

et al. (1981) argue that the two lysozymes have arisen from a common precursor by divergent evolution (and see Selander's observation on the subject in Chapter 3).

In contrast, other enzymes may have developed similar catalytic functions through convergent evolution. Subtilisin and α-chymotrypsin have similar arrangements of amino acid residues in the active sites (Wright, 1972), but the protein folding patterns are quite different, and the positions of amino acid residues involved in catalysis are in different linear sequences. In α-chymotrypsin, the catalytic triad is ordered aspartate 102-histidine 57-serine 195, whereas in subtilisin it is aspartate 32-histidine 64-serine 221. Distinguishing between convergent and divergent evolution is difficult and uncertain, but protein crystallography may provide critical information about the similarities or differences in protein folding, substrate binding, active site residues, and topological alignments.

Not all proteins can be studied by x-ray crystallography, but after one structure has been determined, it becomes possible to attempt to fit the amino acid sequence of another protein onto the three-dimensional structure. Such model building experiments indicate that the liver, yeast, and *Drosophila* alcohol dehydrogenases probably have very similar folding patterns, at least for the nucleotide binding domains. Figure 7 shows one subunit of the liver enzyme with the coenzyme binding domain being the portion of the molecule on the left and the catalytic domain with the two zinc ions on the right. Only two

138

polypeptide chains connect the two domains, and between them there is a large cleft into which the coenzyme and substrate bind. Fitted onto the structure of the liver enzyme is the sequence of yeast enzyme. There are some deletions and insertions, but otherwise the structure is not altered. Although 75 percent of the amino acid residues are exchanged, those that are internal in hydrophobic regions are conserved, similar, or compensated so that the space occupied is the same. The many changes in residues on the surface are readily accommodated. Most amino acids that participate in coenzyme and substrate binding are different, but the changes are conservative so that the essential functions of the side chains are not altered. The substrate binding pocket is somewhat smaller in the yeast enzyme, which explains its narrower substrate specificity. One aspect of the yeast enzyme that remains unclear, however, is how two dimers, as found in the liver enzyme, could interact to form the tetrameric structure of the yeast enzyme (Jörnvall et al., 1978). Similar comparisons have been made for the sequences of several mammalian enzymes and a partial sequence for the enzyme from *Bacillus stearothermophilus* (Eklund et al., 1976).

The sequence of *Drosophila* alcohol dehydrogenase deduced from the structure of the gene has very little sequence homology with the

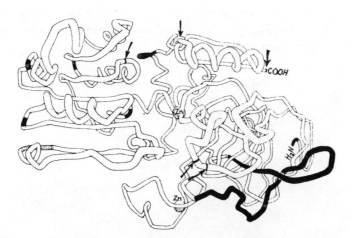

FIGURE 7. Schematic diagram of the tertiary structure of one subunit of liver alcohol dehydrogenase. Regions corresponding to gaps or to segments that are absent in the yeast enzyme are shown in black. Positions corresponding to extra residues in the yeast enzyme are indicated with an arrow. (From Jörnvall et al., 1978.)

yeast or liver enzymes. Nevertheless, prediction of regions of α and β structure suggested that the first 140 residues of the *Drosophila* enzyme had the structure characteristic of the coenzyme binding domain, even though residues 168–318 form this domain in the liver enzyme (Benyajati et al., 1981). After structural alignment, several amino acids are found in equivalent positions, including four residues that are invariant in all dehydrogenase structures: three glycines where a side chain substitution would interfere with folding or coenzyme binding and an aspartic residue that binds a 2′hydroxyl group of the coenzyme. Despite this structural homology, the *Drosophila* enzyme is otherwise very different, since it has 119 fewer amino acids than the liver enzyme and apparently contains no zinc ions.

PROTEIN SUBUNITS AGGREGATE

As shown in Figure 8, liver alcohol dehydrogenase is a dimer of two identical subunits, which are joined together by the coenzyme binding domains and are related by a twofold axis that is in the middle of the molecule. There are two active sites in the cleft regions between the two different domains. Other dehydrogenases form either dimers or tetramers: malate dehydrogenase forms a dimer, whereas lactate- and glyceraldehyde-3-phosphate dehydrogenases form tetramers (Rossmann et al., 1975; Cantor and Schimmel, 1980). It is not clear why these proteins should form the oligomeric structures although it may be noted that sometimes the active sites are formed by amino acid residues from more than one subunit. For instance, with liver alcohol dehydrogenase, the substrate binding pocket is formed by amino acid residues from the catalytic domain of one subunit and the coenzyme binding domain from another subunit.

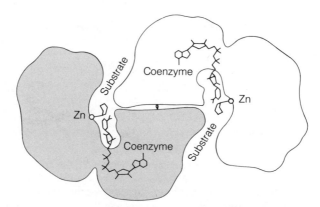

FIGURE 8. Dimeric structure of liver alcohol dehydrogenase, showing binding of the coenzyme and the presumed site of binding of the substrate. Each subunit is identical. (From Brändén and Eklund, 1978.)

140

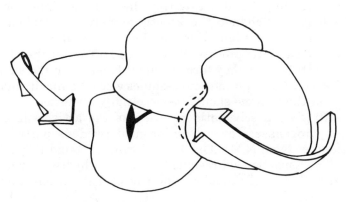

FIGURE 9. A diagram of the overall conformational change of liver alcohol dehydrogenase during the transition from apoenzyme to ternary complex. The twofold axis relating the coenzyme binding domains is indicated by the black tack. (From Eklund and Brändén, 1979.)

PROTEINS ARE FLEXIBLE AND CAN CHANGE CONFORMATION

When liver alcohol dehydrogenase binds coenzyme and a substrate or substrate analog, the structure of the protein changes so that the catalytic domains rotate relative to the coenzyme binding domains (Figure 9). The magnitude of the conformational change is such that amino acid residues located on the surface of the molecule move by as much as 5 Å. There are significant changes in the active site, which is found between the domains. Although the energy of interaction of an enzyme with its substrate is sufficient to alter the equilibrium between conformers, we do not know why the enzyme changes conformation when substrates bind, what triggers the conformational change, or if it is necessary that the conformation change before catalysis occurs. Nevertheless, it seems to be a general phenomenon that enzymes change conformation upon binding substrates.

MANY AMINO ACID RESIDUES FORM THE ACTIVE SITE

Enzymes are quite large structures, but the active sites only occupy a small portion of the total protein. Nevertheless, there are many amino acid residues that participate in forming the active site. Furthermore, there may be as many as two dozen different amino acid

141

residues that participate in binding of a substrate. This means that modification of one amino acid residue by mutation or by chemical modification will in general not destroy the binding of the substrate, even though it may change the kinetic properties significantly. In contrast, it appears that only a few amino acid residues are directly involved in the catalytic process. By direct, I mean that they function as acid/base catalysts or they form covalent bonds with the substrate. If one of these essential amino acid residues were chemically modified, the activity of the enzyme should be essentially abolished.

Figure 10 gives a schematic overview of the active site of liver alcohol dehydrogenase. Several amino acid residues participate in binding the coenzyme NAD. The adenine ring is bound into a hydrophobic pocket. Aspartic acid residue 223 forms a hydrogen bond with the 2'hydroxyl group of the adenosine ribose. This carboxyl group is

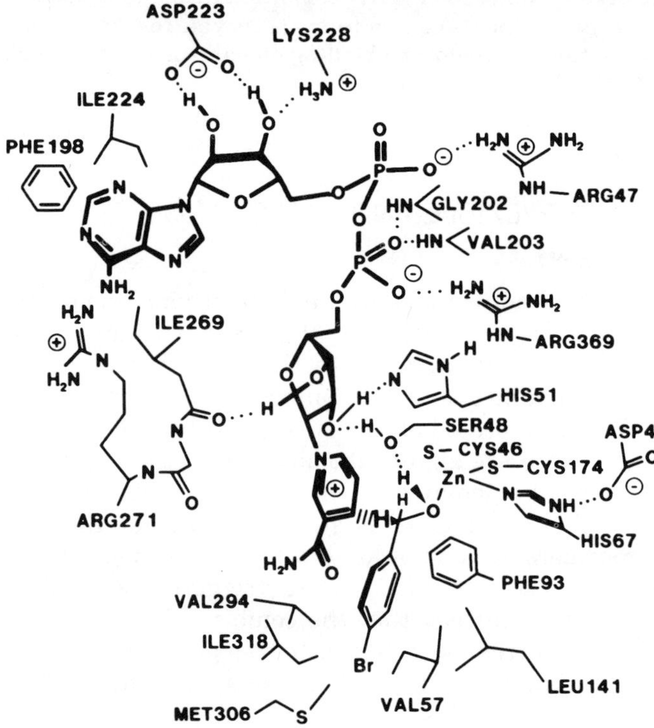

FIGURE 10. Diagram of the active site of liver alcohol dehydrogenase complexed with the coenzyme NAD and the substrate *p*-bromobenzyl alcohol. The pro-*R* hydrogen (the larger hydrogen on the alcohol) is transferred (/// line) to the *re*-face of the nicotinamide ring, which is oriented so that the positively charged nitrogen is toward the front and the carboxamido group is toward the back in this view. (After Eklund et al., 1981; and Plapp et al., 1978.)

invariant in all of the dehydrogenases that have been studied, including the yeast and *Drosophila* alcohol dehydrogenases, and it acts to prevent the binding of nicotinamide adenine dinucleotide phosphate, the coenzyme that has a phosphate ester at the 2' position. The guanidino groups of arginine residues 47 and 369 interact ionically with the pyrophosphate of the NAD. It is interesting that in the yeast enzyme, arginine 47 is replaced by a histidine residue, and this probably results in the weaker binding of the coenzyme by the yeast enzyme. Some evidence for this is that a mutant form of the yeast cytoplasmic enzyme type I that had the histidine residue changed to an arginine residue bound coenzyme three- to fourfold more tightly than did the native enzyme. Nevertheless, even the mutant enzyme does not bind coenzyme as tightly as does the horse liver enzyme (Wills and Jörnvall, 1979a, b).

The substrate p-bromobenzyl alcohol is bound in a hydrophobic pocket, and the alcohol oxygen is directly ligated to the zinc at the active site. This zinc is held in place by two sulfurs from cysteine residues and by one histidine residue. Chemical modification of either of these two sulfurs of the enzyme largely inactivates the protein (e.g., Sogin and Plapp, 1976), but this is probably due to the fact that the charge distribution on the zinc is altered and a bulky group is introduced into the active site of the enzyme where it can interfere with the binding of the alcohol.

In the catalytic mechanism of this enzyme, the pro-R hydrogen from carbon 1 of the alcohol is transferred to carbon 4 on the re-face of the nicotinamide ring (see Figure 10). The proton on the hydroxyl group of the ethanol is also removed so that acetaldehyde is formed. We do not know the exact mechanism for this process, but it is possible that the proton transfer is facilitated by a proton relay system formed by the hydroxyl group of serine 48, the 2' hydroxyl group of the nicotinamide ribose, and the imidazole group of histidine 51, which is on the surface of the protein (Brändén et al., 1975). The coenzyme's hydroxyl group may install the catalytically active species! In the yeast enzyme, serine residue 48 has been replaced by a threonine residue.

Another residue in the active site is lysine residue 228, which has its amino group close to the carboxyl group of aspartate 223 and the pyrophosphate of the coenzyme. One possible function for this amino group is to participate in coenzyme binding. We discovered that chemical modification of this amino group with substituents that retain the positive charge can *increase* the enzyme activity about tenfold (Plapp, 1970; Sogin and Plapp, 1975; Dworschack et al., 1975). A number of

143

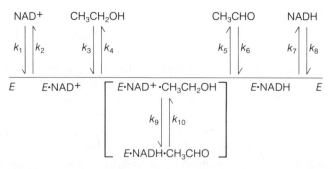

FIGURE 11. The *Ordered Bi Bi* mechanism for liver alcohol dehydrogenase. The binding and release of the two substrates and the two products are ordered.

amino group modifications produce this activation, and the kinetic basis for this activation is interesting. Alcohol dehydrogenase has an *Ordered Bi Bi* mechanism (Figure 11) in which NAD first binds to the enzyme, then the enzyme·NAD complex binds ethanol to form a ternary complex. This ternary complex undergoes the hydrogen transfer reaction to form the enzyme·NADH·acetaldehyde complex. The acetaldehyde dissociates, and then the NADH dissociates, so that the enzyme can participate repeatedly in catalytic turnover. For native horse liver alcohol dehydrogenase, the release of the NADH from the enzyme, with rate constant k_7, is the rate-limiting step in the overall catalytic mechanism. When we chemically modify the amino group of lysine residue 228, however, this step is more than 50 times faster, so that instead of coenzyme release being rate-limiting, the transfer of hydrogen from the ethanol to the NAD, with rate constant k_9, is rate-limiting. Recent x-ray crystallographic studies suggest that a structural explanation for the activation is that modification of the amino group interferes with the conformational change of the protein that occurs when the coenzyme binds. It is interesting that the yeast enzyme also has a lysine at a position corresponding to 228, but when the yeast enzyme is modified, its activity does not increase. This is probably because the yeast enzyme is already about 100 times more active than the liver enzyme so that increasing the rate of coenzyme release does not result in activation. The *Drosophila* enzyme may have a phenylalanine or other uncharged residue at the corresponding position, and it would be interesting to do careful comparative kinetic studies. Although the normal human liver enzyme has a lysine at this position, there is a variant form that has three to five times higher activity and has an alanine substituted for a proline at position 230 (Berger et al., 1974). It appears from inspection of the structure of the protein that a proline in this position would disrupt the α-helix that contains lysine-228 and thereby disturb coenzyme binding.

144

A novel crystallographic method has provided direct evidence for the binding of substrate. Liver alcohol dehydrogenase was crystallized with an active substrate, *p*-bromobenzyl alcohol, from an equilibrium mixture of substrates and products. As shown in Figure 12, the hydroxyl group on the bromobenzyl alcohol is ligated to the zinc, and the *p*-bromobenzyl portion of the substrate is in a hydrophobic pocket which is made up of amino acid residues from two different subunits. Since this pocket is very large it is possible for this enzyme to accommodate a variety of substrates. Although the enzyme's normal physiological substrate may be ethanol, it also can act upon alcohols as large as steroid alcohols. It is interesting that bromobenzyl alcohol appears to be bound in a nonproductive way. The pro-*R* hydrogen of the bromobenzyl alcohol is not pointing toward the nicotinamide ring. Nevertheless, by simply flipping the ring, which would be a very fast process, the hydrogen can be placed in a position where it can be directly transferred to the carbon 4 of the nicotinamide ring, as shown in Figure 10.

Another enzyme that has been extensively studied is lactate dehydrogenase, which is of interest because the distribution of its isoenzymes varies with the tissue and is thought to have physiological significance (Figure 13). Again, there are many different amino acid residues that are involved in binding of the coenzyme. The diagram illustrates the heart type of lactate dehydrogenase, but the muscle

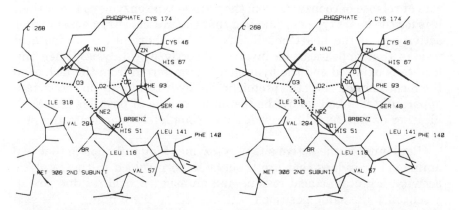

FIGURE 12. Stereoscopic view of the binding of bromobenzyl alcohol into the hydrophobic pocket of liver alcohol dehydrogenase. (From unpublished work by H. Eklund, J.-P. Samama, B. V. Plapp, and C.-I. Brändén.)

145

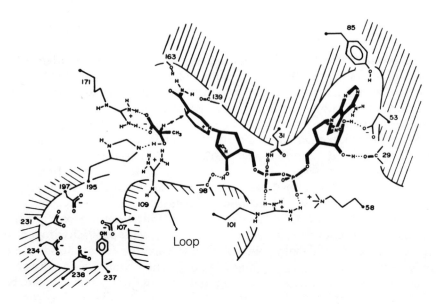

FIGURE 13. Diagram of the principal groups in the active site of lactate dehydrogenase while binding lactate and NAD⁺. (From Eventoff et al., 1977.)

type of enzyme has basically the same overall structure. There are a few differences in the residues in the active sites, but the most important difference is the presence of glutamine 31 in the heart enzyme, which is replaced by an alanine residue in the muscle enzyme. The glutamine residue can form a hydrogen bond with the phosphate of the nicotinamide ribose. An alanine side chain would obviously not form such a hydrogen bond, and this may account for the fact that the rate of release of coenzyme from the muscle type enzyme is faster than it is for the heart type enzyme and that the muscle enzyme has higher activity than the heart enzyme. Note that there is one amino acid residue that appears to be involved directly in catalysis. When the lactate is oxidized and the hydrogen is transferred from the lactate to the nicotinamide ring, the proton on the hydroxyl group on the lactate apparently is taken up by the histidine residue at position 195. Modification of this histidine residue will completely inactivate the protein.

Returning to the active site of lysozyme (Figure 6), notice that two amino acid residues that are thought to be involved in the catalytic activity, a glutamic acid residue and an aspartic acid residue, are in essentially the same positions with respect to the binding of the saccharide in both structures. Small adjustments could put the carboxyl groups into catalytically active positions. There are also some equivalent residues that form hydrogen bonds that hold the acetyl group of the N-acetylglucosamine into the active site.

146

CONCLUSION

The correlation of protein structures determined by x-ray crystallography with their structures and activities in solution has shown that many amino acid residues participate in determining the structure of a protein and its catalytic activity. Although we are now beginning to understand how active sites are constructed, we do not yet have a satisfactory physical-chemical explanation for the catalytic activity of any enzyme. Furthermore, our efforts to correlate the different structures of proteins with their catalytic activities are hindered by the enormous complexity of proteins. Proteins are sensitive to the substitution of amino acid residues, but the effects may be specific or diverse enough to yield manifold results. We still need to study systematically enzymes with altered amino acid residues in order to clarify how each residue is involved in structure and function. Then we may begin to understand the evolutionary origins of proteins.

147

CHAPTER 9

NUCLEOTIDE SEQUENCES AND BACTERIAL EVOLUTION

Irving P. Crawford

Until recently any attempt at a definitive analysis of bacterial evolution would properly have been ignored or greeted with derision by most bacteriologists. Now, however, chiefly as a result of powerful new techniques for manipulating and sequencing nucleic acids, the subject seems no longer to lie in the realm where one person's guess is as good as another's. This survey will begin with a brief review of some aspects of bacterial molecular biology, then consider the twin questions of phylogeny and adaptation among the bacteria.

SOME ASPECTS OF BACTERIAL MOLECULAR BIOLOGY

The long sequence of deoxynucleotides—about 3 million base pairs—composing the genome of an average bacterium has alternating segments of two kinds, those which encode polypeptides and those which do not. Some of those which do not are nevertheless transcribed into RNA. Examples are the genes for ribosomal and transfer RNA molecules as well as short sequences representing the leaders, intercistronic segments and terminal untranslated regions of mRNA molecules. Some considerable but as yet imprecisely known proportion of the DNA is not transcribed at all, however. This includes promoter and operator sites, certain direct or inverted repeats flanking transposable elements, and other regions termed spacer elements whose function is not currently understood. Although some of the sequence constraints on promoters and repressor recognition sites are becoming known (and one should not minimize the importance of these se-

148

quences to bacterial evolution), we know more about the transcribed regions at present, and these represent a larger proportion of the genome.

As an RNA transcript is being translated by ribosomes to make a polypeptide, certain features of the genetic code come into play. It has often been pointed out that the code is redundant. Sixty-one of the 64 codons are used to encode only 20 amino acids. From as few as one to as many as six codons may call for a particular amino acid. Since the commonest amino acids in proteins are those with a greater number of codon possibilities, the weighted average number of codon possibilities per amino acid residue is actually greater than 61 divided by 20. The entire DNA sequence of the five-gene tryptophan operon of *E. coli* has recently been determined in the laboratories of Yanofsky and Platt (Yanofsky et al., 1981). I shall refer to this stretch of DNA often in this chapter. As far as we know, it is typical for its type, a closely regulated group of genes encoding a biosynthetic pathway, expressed intermittently and at a moderate level. The entire sequenced region, including 270 base pairs upstream from the start of transcription and 330 base pairs downstream from the end of the last gene, is 7,288 base pairs in length. Its five structural genes encode a set of rather typical enzymes, the largest polypeptide being 531 amino acids in length and the smallest being 268.

There are more than 2,000 working codons in this stretch of DNA. Certain features of the amino acid content of the proteins are shown in Table 1, where the amino acids are grouped according to the number of possible codon choices. The only striking deviation in frequency is seen in the gross underutilization of tryptophan, perhaps understandable in a group of enzymes whose synthesis must be maximized when that amino acid becomes scarce. Even neglecting tryptophan, however, the trend for commonly employed amino acids to have a large number of possible codons is clear from the utilization index, which is significantly greater than one for the amino acids with four or six codons and significantly less than one for those with one or two codons. This results in an average of 3.6 codon possibilities for each position in the polypeptide chain. Clearly this freedom of codon choice permits considerable variation in DNA sequence even if the amino acid sequence is held constant. As inspection of the code will show, most of this variability resides in the third codon position, yielding a predictable pattern of potential base substitutions three (or a multiple of three) bases apart. As we will see later, this pattern has been observed in comparisons of the *trp* genes of closely related bacteria.

Of course, it is also true that amino acid substitutions can occur at

149

TABLE 1. Amino acid usage in the *E. coli* tryptophan operon.

Number of synonymous codons	Amino acid & number of times it is used	Utilization index[*]
6	Arg 116 Leu 248 Ser 121	1.50
4	Ala 248 Gly 167 Pro 111 Thr 97 Val 137	1.43
3	Ile 112	1.04
2	Asn 80 Asp 116 Cys 32 Gln 103 Glu 134 His 63 Lys 78 Phe 76 Tyr 62	0.76
1	Met 51 Trp 5	0.47

[*] Tryptophan usage is excluded from this index (see text), which reflects the relative frequency of occurrence of the amino acids in each group.

many positions in the polypeptide chain without detectable loss of function. In this way, as well as with certain synonymous codon substitutions for arginine, leucine, and serine, changes in the first and second codon positions can occur without affecting the phenotype. All these nucleotide substitutions constitute the raw material for a kind of DNA evolution not associated with obvious selectable changes in function.

CODON USAGE

Again referring to the *E. coli trp* operon, it is apparent from Table 2 that even though all of the 61 sense codons are employed at least once, there is a wide range of utilization. One of the arginine codons (AGG) is used only once and an isoleucine codon (ATA) only twice, while a number of other codons are used 50 to 100 times and one particular leucine codon (CTG) is used 133 times. There is no immediately obvious pattern of codon preference among the sets having four or six synonymous codons, other than a slight tendency to avoid codons ending in A. In most cases all possible codons receive reasonably heavy use, however. For amino acids having two synonymous codons, the greatest disproportion is in the codon usage for lysine, where the codon ending in A is in fact favored. Glutamate shows the same tendency.

The pattern of codon usage seen for the *trp* operon is similar to that of a number of other *E. coli* genes expressed at a moderate level. It differs from that seen for genes expressed at a very high level, such as the ones encoding ribosomal proteins (Post et al., 1977) and some structural proteins in the outer cell-wall layer (Nakamura and Inouye,

150

1980) where not all codons are used. Recently Grantham et al. (1981) proposed that highly expressed genes show an avoidance of codons with extremely high or low binding energies for the cognate tRNA. Apparently this is one of the strategies employed by *E. coli* to ensure a high rate of translation.

Genes of low to intermediate expression, on the other hand, have greater freedom of codon usage, and it is here one sees the organism's overall G+C content reflected in the codon choices, especially in the third codon position. We lack data for the entire *trp* operon in any organism other than *E. coli*, but have sequence data for some of the *trp* genes of *Salmonella typhimurium, Klebsiella aerogenes,* and *Serratia marcescens,* all rather close relatives of *E. coli* and classified together with it in the enteric bacterial family. Table 3 shows the proportion of G+C bases in the entire genome, in the first and second and in the third codon positions in four segments that have been compared. Variation in overall G+C content is reflected slightly in the first and second codon positions, but decisively in the third position. The organisms studied do not cover the extremes of G+C content

TABLE 2. Codon usage in the *E. coli* tryptophan operon.

Arginine		Leucine		Serine		Alanine		Glycine		Proline	
CGT	38	TTA	30	TCT	17	GCT	38	GGT	62	CCT	13
CGC	66	TTG	30	TCC	21	GCC	86	GGC	68	CCC	16
CGA	5	CTT	20	TCA	12	GCA	43	GGA	15	CCA	19
CGG	3	CTC	24	TCG	21	GCG	91	GGG	22	CCG	63
AGA	3	CTA	11	AGT	17						
AGG	1	CTG	133	AGC	33						

Threonine		Valine		Isoleucine		Asparagine		Aspartate		Cysteine	
ACT	12	GTT	33	ATT	67	AAT	32	GAT	73	TGT	14
ACC	50	GTC	27	ATC	43	AAC	48	GAC	43	TGC	18
ACA	13	GTA	18	ATA	2						
ACG	22	GTG	59								

Glutamine		Glutamate		Histidine		Lysine		Phenylalanine		Tyrosine	
CAA	38	GAA	96	CAT	32	AAA	64	TTT	39	TAT	39
CAG	65	GAG	38	CAC	31	AAG	14	TTC	37	TAC	23

151

TABLE 3. G+C content of portions of the tryptophan operon (%).

	E. coli	*S. typhimurium*	*K. aerogenes*	*S. marcescens*
Entire genome	51	52	56	59
trpA (269–270 codons)				
1st and 2nd positions	53	54	56	
3rd position	56	63	83	
trpB (398 codons)				
1st and 2nd positions	52	53		
3rd position	60	64		
trp(G)D (194 codons)				
1st and 2nd positions	52	54		56
3rd position	60	62		82
trpE (521 codons)				
1st and 2nd positions	54	56		
3rd position	55	62		

Sources: Entire genome, Sober (1970); *trpA,* Nichols et al. (1981); *trpB,* Crawford et al. (1980); *(G)* segment of *trpD,* Nichols et al. (1980); *trpE,* Yanofsky and van Cleemput, 1982.

among even the close relatives of *E. coli.* As more DNA sequences become available, we can expect to see wide fluctuations in the proportion of G+C bases found in the third codon position in genes whose expression rate is normally low or intermediate.

Other than reflecting the G+C content, codon usage in moderately expressed genes generally approximates the availability of the cognate species of transfer RNA. This still allows enough flexibility in codon use so that *E. coli* and *S. typhimurium* sequences can differ appreciably even for proteins whose amino acid sequence is highly conserved. Thus, for three of the *trp* genes the divergence in base sequence is: *trpA,* 25 percent divergent but 72 percent of the changes are to synonymous codons; *trpB,* 16 percent divergent but 90 percent of the changes synonymous; *trpE,* 20 percent divergent but 70 percent of the changes synonymous. Together these three genes account for about one-half of the coding sequences in the operon. Most of the synonymous changes are in the third codon position, of course, giving rise to a characteristic pattern resembling random base substitutions three (or a multiple of three) bases apart (Crawford et al., 1980).

The DNA sequence differences described so far are chiefly ones that do not involve an alteration of the amino acid sequence. Obviously, as a result of mutation and selection, additional modification of DNA sequence has taken place during the evolution of bacteria. Let us first consider the appearance of new catalytic activities.

When the *lacZ* gene specifying β-galactosidase is deleted from the

152

E. coli chromosome, the cell is unable to grow on lactose. As a result of rare genetic events at a different location on the chromosome, however, a new lactose-splitting activity can appear, and the cell again becomes lactose positive (Warren, 1972; Campbell et al., 1973; Hall and Hartl, 1974). The locus for the novel enzyme has been termed *ebgA* for evolved β-galactosidase (Campbell et al., 1973; Hall and Hartl, 1974). Although operator-constitutive mutants for *ebgA* are sometimes found in these cells, in fact a more common event is that a regulatory mutation making the *ebgA* enzyme inducible by lactose has occurred in company with a structural alteration of the enzyme making lactose a substrate (Hall, 1977). This differs somewhat from the acquisition of xylitol utilization by *K. aerogenes* (Wu et al., 1968), where an existing ribitol dehydrogenase having weak activity toward xylitol as a side reaction first becomes constitutive as a result of a regulatory mutation and later is modified by structural gene mutations to permit faster growth. A similar situation has been uncovered by Clarke (1974) in the utilization of formamide and butyramide by *Pseudomonas aeruginosa*. Hall (1978) has shown that a variety of mutations can occur in *ebgR,* the gene producing the repressor for *ebgA*. Some of these broaden further the specificity of induction so the system responds to such unusual substances as lactulose (fructose-β-D-galactoside), methyl-galactoside, and galactose-arabinoside. Thus it is clear that both inducer recognition and enzyme specificity can be altered by mutation in this system. Similar events have doubtless occurred many times in bacterial evolution. In the most popular scenario, they are preceded by duplication of the gene or genes in question so the original function of the locus need not be lost.

In addition to mutations creating what are essentially new activities for old genes, we find evidence for the fusion of old genes of related function to form new complex enzymes. One of the best known cases involves the first two enzymes of the tryptophan pathway in enteric bacteria. In most nonenterics, and in some enterics like *Serratia marcescens,* this enzyme has a large and a small subunit (Zalkin, 1973). The large subunit's gene has remained separate in all bacteria studied to date, but in some enterics the gene for the small subunit has been fused with the gene immediately downstream from it, the one for the second enzyme in the pathway (Henderson and Zalkin, 1971; Yanofsky et al., 1971). This results in *E. coli* and its closest relatives having a new complex enzyme consisting of the large subunit of the first enzyme bound noncovalently to the fused polypeptide made from the small subunit of the first enzyme and the second enzyme. This has certain enzymatic consequences, such as bringing the active

153

sites of these two enzymes into proximity in a fixed relationship to each other (Zalkin, 1973). More to the point of the present topic, we can reconstruct the changes in DNA sequence required to accomplish the fusion. Figure 1 shows the sequences near the point of fusion (Miozzari and Yanofsky, 1979) and outlines one scenario leading from the separate genes of *S. marcescens* to the fused gene situation in *E. coli*. At the first level, after alteration of a nonsense codon to sense and deletion of a single base pair to establish the correct reading frame, the genes are fused. The second level involves missense changes in three codons to eliminate the ribosome binding site for the second gene. The third level involves other sequence differences in this region having less dramatic or inconsequential effects. We do not know the order in which these changes occurred, or whether in fact the third-level changes occurred in the *Serratia* or the *Escherichia* line of descent.

Other examples of gene fusion are known in this short pathway. Two of the central enzymes of the pathway are the products of separate genes in other bacteria but are fused in the enteric group (Yanofsky et al., 1971; Crawford, 1975, 1980). In this case, the separate genes are usually not even linked, so the fusion event must have involved transposition, deletion, and the elimination of a stop codon.

Transpositions, of course, are now thought to occur at least as readily in bacteria as elsewhere. Some of their effects on the chromosomal distribution of the genes of the tryptophan pathway are shown in Figure 2. Here the *trpG-trpD* fusion in *E. coli* and the *trpC-trpF* fusion in all enteric bacteria are shown; *trpR* is the unlinked regulatory gene for the five- or six-gene operon. Two other gram-negative bacterial groups have different and nonidentical patterns,

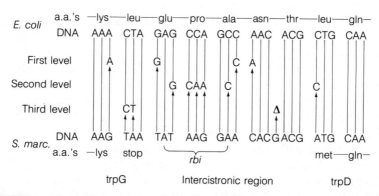

FIGURE 1. Hypothetical scheme for the establishment of the enteric bacterial *trpG-trpD* gene fusion. The portion of the *S. marcescens* intercistronic region designated *rbi* is the ribosome binding and initiation region. (After Miozzari and Yanofsky, 1979.)

154

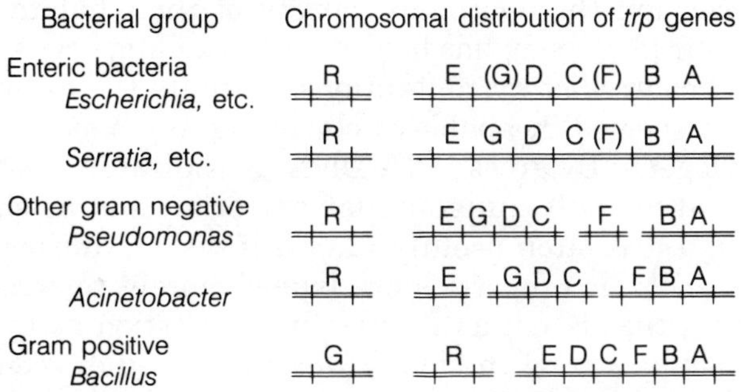

Bacterial group   Chromosomal distribution of *trp* genes

FIGURE 2. Tryptophan gene distributions in bacteria. Analogous genes are given the same letter designation throughout, but the *trp* prefix for all gene symbols has been omitted for simplicity. Genes *E* and *G* encode the first pathway enzyme; *D, F,* and *C* the second, third, and fourth enzymes; and *B* and *A* the β and α subunits of the fifth enzyme, tryptophan synthase. In *Acinetobacter* the gene order in the three-gene clusters is not certain; the most probable order is shown. (After Crawford, 1980.)

with the structural genes for the pathway residing in three unlinked locations (Crawford, 1975). *Bacillus subtilis,* a gram-positive organism, is different still (Hoch et al., 1969; Kane et al., 1972), and the figure could be enlarged to show even more variations. That each of these situations requires attendant DNA-sequence changes is immediately apparent from the necessity for new promoters and terminators for transcription. Moreover, in some cases transposition of a gene is accompanied by a complete change in its regulatory apparatus. Thus, in *Pseudomonas putida* and *Pseudomonas aeruginosa* the *A, B,* and *F* genes are not at all responsive to the repressor produced by the *trpR* gene (Maurer and Crawford, 1971; Calhoun et al., 1973). Instead, the *F* gene has been left unregulated, while an entirely novel apparatus has been developed for the *A* and *B* genes, making them inducible by their substrate rather than repressible by tryptophan (Crawford and Gunsalus, 1966; Manch and Crawford, 1981).

    I will summarize this brief consideration of the molecular genetics of bacteria by noting that the sequence of the bulk of the DNA, which encodes proteins, can and does change very significantly without affecting the function of these proteins, and that the evolution of new genes from preexisting ones, the rearrangement and fusion of old genes, and the modification of regulatory devices all contribute further to diversification of the DNA.

155

It is little wonder, considering the small size and restricted morphology of bacteria and the enormous variety of their habitats and life styles, that their phylogeny has been a long-standing arena of conflicting views and contradictory contentions. Initially the examination of bacterial DNA seemed incapable of clarifying any major questions of bacterial phylogeny. By gross DNA-DNA hybridization, or by mRNA-DNA hybridization, only quite similar bacteria could be shown to be related. Though it is often useful to know if two organisms sharing a number of phenotypic properties are indeed closely related in a phylogenetic sense, that is not as interesting a question as the phylogenetic relationships among the many weakly related or seemingly unrelated bacterial groups. Within the past five years or so, however, it has become apparent that certain segments of the bacterial genome are much better conserved than the majority of it. By paying attention only to these highly conserved areas, a fairly unambiguous phylogenetic tree can be derived. This tree is still in the early stages of definition, and its derivation has already been described by Selander in Chapter 3, so I will not consider it in great detail here. Suffice to say a wide gap separates one group of bacteria—the methane producers, halophiles, and certain wall-less forms able to tolerate high temperatures and low pH—from the great mass of "normal" bacteria, including the chemoautotrophs, photosynthetic bacteria, and blue-green bacteria. The former group has been designated the archaebacteria by Woese and Fox (1977). The latter can be called the eubacteria, redefined to include the prominent group formerly known as the blue-green algae. Incidentally, the term archaebacteria should not be taken to imply that organisms of this type greatly preceded other life forms on earth. All that can be claimed is that these bacteria diverged from the eubacteria a very long time ago. Figure 3 briefly summarizes and diagrams some of these results; it has been redrawn from a recent summary by Woese and his collaborators (Fox et al., 1980).

At the bottom of the figure are the archaebacteria, while the eubacterial radiation at the top, somewhat simplified, shows the following: blue-green bacteria giving rise to chloroplasts; the green and the purple photosynthetic bacteria, the latter giving rise to many of the common gram-negative types; a gram-positive branch generating the spore-formers and the actinomycetes (note that the wall-less *Mycoplasma* seem to have a home among the *Clostridia* and the lactic acid bacteria which are not shown); and *Micrococcus radiodurans* and the spirochaetes, each occupying separate branches. This is certainly a different phylogeny from the mainly intuitive ones proposed earlier, and the question deserves to be put—is it any more likely to last and to become accepted as valid than its predecessors?

156

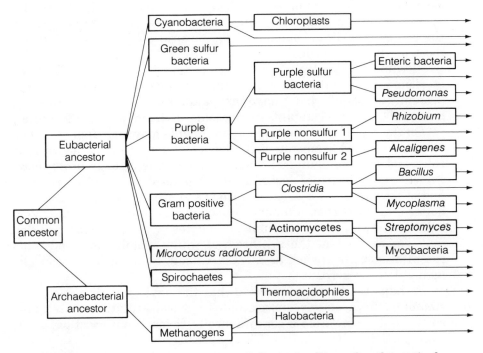

FIGURE 3. Schematic representation of the major lines of prokaryotic descent. (After Fox et al., 1980.)

It is based on the sequence of a small segment of the genome, that determining the 16S ribosomal RNA. It should be noted at the outset that neither in Woese's approach (Fox et al., 1980) nor in rRNA-DNA hybridization (Pace and Campbell, 1971; Johnson and Francis, 1975) is an actual sequence of nucleotides determined. In one case similarities in the complement of oligonucleotides greater than six bases long are estimated; in the other case the relative strength of hybridization between ribosomal RNA and the complementary strand of DNA from the chromosome is measured. Where comparison is possible, these two metrics seem to give similar results. Most observers, probably even the most enthusiastic proponents of each method I have mentioned, would prefer to have a precise nucleotide sequence of the ribosomal RNA. At the present rate of development of sequencing techniques, it should not be more than a few years before it will be feasible to determine the sequence precisely.

Why are these RNA molecules so strongly conserved during evo-

157

lution? The ribosome is a complex apparatus, consisting of a large number of proteins bound to and interacting with the ribosomal RNA. There are a large number of ribosomes in the cell. Even so, this apparatus must be capable of functioning at a nearly optimal rate for the cell to be able to compete in the natural environment. Probably most of the nucleotides in ribosomal RNA have a precise function (Noller and Woese, 1981), and most changes in sequence will be deleterious—more so than for a nucleotide sequence encoding an average protein. If most of what I have just said applies, there is a rationale behind the observed stability of ribosomal RNA in bacterial evolution.

On the other hand, these same arguments may well apply to other, less well-studied segments of the bacterial genome that play crucial roles and may be similarly conserved. There is already preliminary evidence that the genes encoding ribosomal proteins, RNA polymerase and other molecules concerned with the synthesis of proteins may be better conserved than ordinary anabolic and catabolic enzymes. Very little is known about molecular aggregates or complex enzymes that may be involved in the synthesis of walls, membranes, and other essential structures such as the DNA itself. Some of these may also be highly conserved.

Perhaps the best confirmation that the phylogeny discerned from ribosomal RNA closely approximates natural evolutionary relationships would come from a concordance of several conserved but unrelated portions of the genome. Theory predicts that this will occur, and, if so, it will provide confidence similar to that felt when the examination of cytochrome *c* sequences in metazoans generated the same phylogeny that had been discerned by morphological, paleontological, and physiological studies (Fitch, 1976*b*).

Incidentally, the hope that a similar determination of the amino acid sequence of a simple, small protein might provide prokaryotic phylogenetic information seems to have faded, mainly because no really universal, paradigmatic protein present throughout the breadth of the prokaryotic kingdom has been found. A protein measuring stick suitable for examining prokaryotic evolution may appear eventually; perhaps one way to seek it out would be to search by hybridization across a wide phylogenetic range for a well-conserved, gene-sized DNA segment.

Note that throughout this discussion of bacterial phylogeny I have taken no notice of the dispute about whether all the differences in the protein and DNA sequences of these organisms have some selective value or whether many or most of them are selectively neutral. It seems logical that some third-position changes from one to another commonly used codon for an amino acid such as alanine would have no selective effect, and it is true that changes of this sort appear to occur quite at random throughout a coding sequence such as the *trp* operon, but in fact the critical experiments to prove that these have no selective value seem not to have been done, though available techniques for gene transfer between various prokaryotic species would seem to make it possible to do them.

BACTERIAL ADAPTATION

As with all living things, bacteria have the capacity to respond to changes in the environment and to invade diverse habitats. Bacterial phylogeny is obviously the result of a long term, more or less permanent genomic response to habitat opportunities. Short-term, reversible adaptive responses have also been developed, of course, encompassing all the known gene regulatory mechanisms. Recently a novel method of regulating gene expression by inversion of a small segment of the chromosome has been found in some enteric bacteria and their phages (Simon et al., 1980). This is responsible for phase variation in *Salmonella* species, the alternation between two flagellar types after a certain number of generations. This event is completely reversible; inversion of the tiny controlling segment produces no permanent genomic change.

We are not without examples of genetic events that are adaptive and that do produce a more or less permanent change in the chromosome, however. Some have already been mentioned, particularly the appearance of multifunctional polypeptides through gene fusion. This, along with the deletion of unused coding segments and the evolution of new enzymatic specificities, must at times offer cells advantages under altered environmental circumstances. There are also examples of transposons inserting in novel places in the genome and being moved from cell to cell by plasmids and bacteriophages. Examples of selectable antibiotic resistance markers borne by transposons are by now commonplace (Kleckner, 1977), but there are also other instances of enzymes conferring a selective advantage in certain habitats being transported from cell to cell and integrated in novel places via transposition (Calos and Miller, 1980). The entire question of position effects on the bacterial chromosome is opened up by these possibilities.

It is also clear that in many bacterial species a great deal of useful genetic information is borne on plasmids, some of them rather large. Some examples are the degradative plasmids in fluorescent pseudomonads (Jacoby and Shapiro, 1977; Chatterjee et al., 1981), the tumor-inducing plasmids in *Agrobacterium* (Sciaky et al., 1978), and the nitrogen fixation genes in some species of *Rhizobium* (Nuti et al., 1979). The advantage the organism obtains by having these units of

159

infectious heredity can be debated; they may afford the species a greater adaptability and shorter response time, but again the decisive experiments to test this have not been done.

One fact that has become obvious in the study of bacteriophages is that they can exhibit an astonishing parsimony, by compressing a great many functions into a very small amount of DNA. We know there can be multiple uses of a coding sequence, either in the same or different reading frames, and the amount and time of appearance of each protein product can be rigorously controlled. Similar economy in the use of bacterial DNA has not yet been reported, but we must bear in mind that if it served an adaptive function it would be potentially available.

HIGHLY VARIABLE TRAITS

It is an old observation that certain phenotypic traits of bacteria seem to be more variable than others. Many of these are associated with the outer surface of the cell and can be detected by immunological tests directed against the intact organism. The Kauffman-White scheme for classifying *Salmonella* (described by Le Minor and Rohde, 1974) is an example of an enormous proliferation of "species" all quite similar genetically but differing in surface immunological determinants. It seems possible that many bacteria, if studied in similar detail, would be found capable of giving rise to a similar profusion of types based on small surface changes. The polysaccharide and protein molecules that have been evolved for this location may in fact have been designed in part to facilitate this sort of variation (Stocker and Mäkelä, 1978; Diderichsen, 1980). It seems attractive at first to suppose that this variability represents an adaptive response on the part of a symbiont or pathogen to its host's immunological defense mechanism, but a little reflection leads one to the conclusion that evolution of the bacterial cell wall must have greatly preceded the development of immune defenses in vertebrates. It is likely that bacterial viruses go back almost as far as bacteria, however, and the same structures recognized by antibodies often serve as attachment sites for viruses. It is a reasonable hypothesis that the variability of surface features evolved as an adaptation to pervasive bacteriophages, and that an uneasy equilibrium has long existed between the ability of a bacterium to change its surface and that of a virus to modify its attachment organ in order to recognize the new one.

Another feature of the bacterial cell that may have adaptive value and seems to be quite variable is the occurrence of restriction and modification enzymes. As is well known, restriction enzymes recognize a four-, five-, or six-base sequence in double-stranded DNA and cleave

160

it; modification enzymes see the same site and protect it, often by methylating one of the bases. The presence of both activities allows a bacterium to recognize and cleave invading foreign DNA while ignoring its own. Probably this too evolved as a defense against attack by unwanted phages. Whatever its origin, it seems reasonable that protection from DNA that comes from, or has been replicated in, close relatives is needed most. Thus different strains of the same species have been shown, in some instances at least, to possess restriction and modification enzymes with different recognition sequences (Roberts, 1981). I am not aware of an investigation of this phenomenon on anything like the scale of the Kauffman-White scheme, but it would seem feasible to ask if the specificity of these internal enzymes is a trait as variable as the antigenic nature of the cell surface.

We have been impressed by the high rate of third-base, synonymous codon changes when comparing the DNA of rather closely related enteric bacteria, a rate approaching four to five events for every base change leading to an amino acid change in the *trpE, trpA,* and *trpB* genes of *S. typhimurium* and *E. coli* (Yanofsky and van Cleemput, 1982; Nichols and Yanofsky, 1979; Crawford et al., 1980). We have wondered if this might be an adaptive event leading to ease of recognition of foreign DNA, and perhaps failure to recombine with it, through the accumulated sequence differences. It is difficult at present to imagine an acceptable mechanism for performing this change, or to deduce the manner in which the cell recognizes its value and encourages it once it has occurred.

Before leaving the subject of bacterial adaptation, I want to look at the other side of the coin and correct the impression I may have given that all bacterial DNA sequences outside a few highly conserved ones like the ribosomal RNA genes are so variable that little of evolutionary significance can be learned from determining them. When an average protein is being encoded, there is good reason to believe that, in order for it to function properly, portions of its amino acid sequence will be held nearly constant over a long period of evolution. As an indication of this, Figure 4 presents a figure from a recent paper by Zalkin and Yanofsky (1982).

In yeasts and molds the enzyme tryptophan synthase is known to be a dimer composed of a single type of polypeptide chain (Manney et al., 1969; Matchett and DeMoss, 1975), though it carries out enzymatic reactions and side reactions identical to those performed by bacterial tryptophan synthase, which is made up of two α and two β chains. From the size of the single yeast polypeptide and some properties of nonsense mutants (Manney et al., 1969), it had been hypothesized to

161

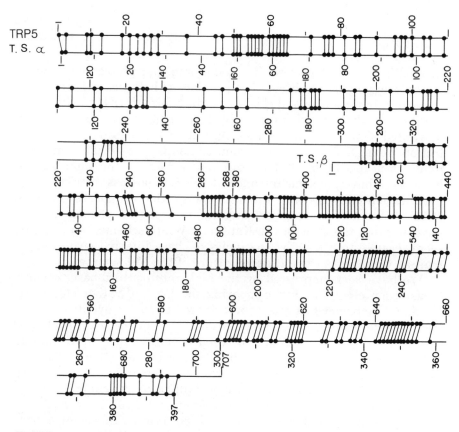

FIGURE 4. Comparison of *Saccharomyces cerevisiae* tryptophan synthase with *E. coli* α and β subunits of the same enzyme. Protein chains are drawn to scale, yeast above and *E. coli* below. Identical amino acids are indicated by dots and connecting lines. A vertical line indicates a perfect match. A slanted connecting line indicates that a one-codon insertion or deletion is required to obtain the match. (From Zalkin and Yanofsky, 1982.)

be a fusion polypeptide having the α homolog N-terminal and the β homolog C-terminal. By decoding the amino acid sequence from the DNA sequence of *tryp-5*, the gene for tryptophan synthase in yeast, Zalkin and Yanofsky (1982) showed that this supposition was correct. Amazingly, 50 percent of the amino acid residues in the β portion are identical to ones in *E. coli,* and 30 percent of the residues in the α portion are conserved. About a dozen of these conserved residues have been designated as essential for catalysis in the active site, but most seem necessary to maintain the domain structure and the essential three-dimensional polypeptide framework. When it is recalled that all phylogenies place eukaryotic nuclear genes and prokaryotic genes about as far apart as possible, this conservation of amino acid sequence in the face of millions of years of nucleotide sequence divergence is indeed a strong argument for the power of selection in stably maintaining those features of the polypeptide needed for enzymatic activity.

SUMMARY

The DNA of bacteria consists of a smaller untranscribed part, about which little is known in an evolutionary sense, and a larger transcribed part. At least one transcribed segment, the one giving rise to ribosomal RNA, is highly conserved (very likely there are others also) and offers the best current prospect for obtaining a meaningful phylogeny of the prokaryotes. Most of the rest of the transcribed DNA, which forms the bulk of it, encodes proteins and their regulatory apparatus; this part shows a very high rate of DNA sequence divergence. Between close relatives, these are mainly changes in the third base of codons and a few other changes whose effect on the phenotype is minimal. As more distantly related forms are examined, more drastic changes involving numerous amino acid substitutions, transpositions, gene fusions, and activity modifications are seen, presumably reflecting changes in the habitat and metabolic mode of life of the organisms. Even in very distantly related types like yeast and *E. coli,* however, certain constraints on the function of proteins which perform identical jobs require amino acid sequence similarities which may reach 50 percent of the residues in favorable cases. Presumably a careful consideration of several conserved DNA and protein segments will in the foreseeable future result in a consensus concerning the paths of bacterial evolution. Then we can use the many favorable features prokaryotes exhibit, such as compactness of genome and rapid generation time, to unravel the mechanisms creating and regulating their evolutionary differences.

CHAPTER 10

THE IMPACT OF MODERN GENETICS ON EVOLUTIONARY THEORY

Tim Hunkapiller, Henry Huang, Leroy Hood,
and John H. Campbell

Evolution is a tangible and dynamic aspect of biological communities. However, while evidence of its occurrence is overwhelming, its precise nature is much more obscure. The pioneering work of Haldane, Fisher, Wright, and others in the 1920s and 1930s laid the theoretical framework for population genetics. From this foundation, Huxley, in the early 1940s, developed the first widely accepted model for evolutionary change and the forces involved—the "modern synthesis" or Neodarwinism. This theory is essentially a marriage of Mendelian genetics and classic Darwinian selection and has been a cornerstone of modern biological thought. Today, however, the theory of evolution as embodied by Neodarwinism is a theory in some turmoil. There are phyletic gradualists and punctuated equilibrists. There are strict selectionists and just as strict neutralists. There are altruistic genes that turn out to be selfish. There are even reports that acquired traits can be inherited. Unfortunately, classic paleontological and taxonomic approaches to these and other issues and problems have yielded results too equivocal for general consensus. Even the refinement of taxonomy to the biochemical level has stirred as much controversy as it has shed light.

In just the last few years, exciting technological advances in the field of molecular genetics have been responsible for a revolutionary new view of the structure and organization of genetic information.

164

Likewise, this new perspective has suggested many intriguing and novel means of genetic variation. It is our belief that this new model of the genome imposes both constraints and potentialities on any new synthesis of evolutionary theory and should aid in resolving many of the outstanding issues of debate. In this chapter we will present a brief outline of several of the phenomena responsible for this new genetic model, in order to convey something of the current view of the complex nature of genetic elements, their functional and structural organization within the genome, and the modes of physiological variation of germline information.

The genetic elements that we will discuss represent, by themselves and as components, different levels of organizational complexity that generate hierarchical strategies of function and fitness. Variation, it seems, is an integral part of the dynamic equilibrium between these systems. The rate and style of evolution then depends not only on the external forces exerted upon organisms but also upon the nature of variation prescribed by their genetic organization.

We will start our discussion with a brief description of some particular genomic structural elements and the implications they have for genetic variation. We will then describe examples of the higher level functional organization of these components and the possibility they provide for integrated genetic change.

SPLIT GENES—EXONS AND INTRONS

Perhaps no observation has come as such a great surprise to molecular biologists as the fact that eukaryotic genes are split into a series of alternating peptide-coding regions (exons) and intervening sequences (introns) that are transcribed together as a single, high molecular RNA species. Known exon numbers range from two to as many as 52 for α-collagen (for review, see Abelson, 1979). Subsequent enzymatic excision of the noncoding information from the RNA transcript is required to generate a functional messenger RNA in which all coding information is contiguous and translatable. This process is called RNA splicing (Figure 1).

The discovery of split genes was followed by the corollary observation that often functionally discrete domains of particular proteins are coded for by individual exons (Blake, 1981). The simplest and perhaps clearest example of this phenomenon is seen in the antibody system where discrete exons encode homologous structural domains of the antibody molecule (Calame et al., 1980; Sakano et al., 1979) (Fig-

165

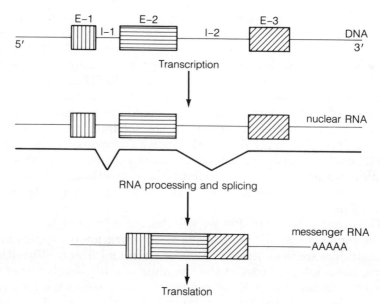

FIGURE 1. A model of RNA splicing. The original RNA transcript of a split gene includes both coding (exons or E) and noncoding (introns or I) sequences. RNA processing, which includes the splicing together of all exons and the subsequent loss of introns, produces a mature, translatable messenger RNA. Non-protein coding RNAs (transfer RNAs, ribosomal RNAs) undergo a similar RNA splicing process.

ure 2). In this context, the evolution of many multidomain proteins, like antibodies, can readily be explained by the duplication of a single primordial exon and the subsequent divergence of the homologs. Thus, exons encode evolutionary as well as functional units in proteins. The duplication of functional domains without exon splicing would require the far less likely event of precisely contiguous and in-phase alignment of the duplicate and original coding sequence. Even more striking are the evolutionary implications of the rearrangement of exons referred to as "exon shuffling" (Doolittle, 1978; Darnell, 1978; Gilbert, 1978; Crick, 1979). When genes are composed of multiple discrete exons that encode distinct functional or structural peptide domains, new combinations of exons can generate novel synergistic combinations of these domains. Any intra- or intergenic recombinational event or mutation in splice-site signal sequence can generate unique permutations of this kind. Isolation of the coding information as exons surrounded by nontranslated DNA sequence helps maintain the integrity of genetic information during such events and insures a proper reading frame alignment. This evolutionary strategy removes the requisite of rein-venting the wheel or rather reevolving certain structural and/or func-

166

tional strategies before they are employed within significantly new contexts. For example, recombination of the exon encoding the membrane-anchoring domain of one gene with a second gene whose product previously was secreted drastically alters the functional context of the second protein. Exon shuffling would allow significant evolution of proteins to proceed in major steps rather than through a series of intermediates reflecting minor incremental changes. Since the domains involved in exon shuffling would have already experienced selection for their structural and functional viability, selection on new combinations will need to operate less on this aspect and more on the functional relationship of the new product and the cell producing it. This view does not imply that eukaryote genomes are organized to encourage such evolution, but rather that the evolution is a fortuitous by-product of an arrangement that serves other primary functions.

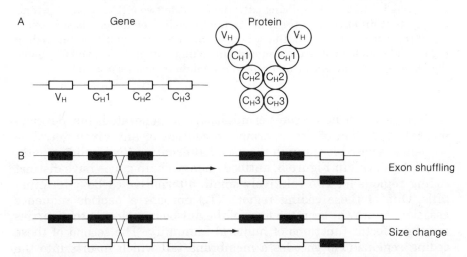

FIGURE 2. A. The correlation of immunoglobulin exons and globular domains. The DNA organization of an immunoglobulin heavy chain gene and domain structure of the corresponding heavy chain polypeptide are represented. The boxes represent exons, and the lines represent flanking sequences or introns for the gene. The circles represent discrete globular domains for the protein (the light chains are omitted for clarity). There is a perfect correlation between the exons and the corresponding globular domains. B. Exon shuffling. Classical genetic phenomena can result in the rearrangement or "shuffling" of exon information. Shown are possible rearrangement events involving immunoglobulin heavy chain constant region genes that could result in new exon arrangements and, subsequently, new synergistic associations of functionally and/or structurally discrete domains.

167

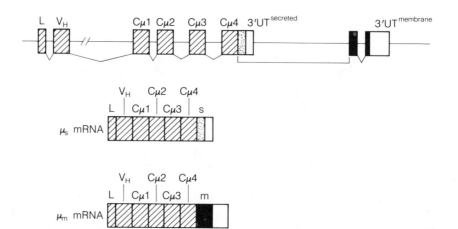

FIGURE 3. Antibody molecules are expressed in either the secreted (s) or the membrane-bound (m) form, depending on the RNA splicing and processing pathways employed after transcription of the genes. The exons (boxes) and introns (lines) of the μ heavy chain gene (encoding antibodies of the IgM class) are depicted. Alternative splicing patterns result in two mRNAs, $\mu_{membrane}$ and $\mu_{secreted}$, that differ only at the 3′ (C-terminal) coding end. Open boxes denote the 3′ untranslated regions. Black boxes represent the membrane tail coding region and stippled boxes the secreted tail-coding region. L, V_H, and C_μ denote the leader-, variable-gene-, and constant-domain exons, respectively.

Not only can new exon permutations be generated, but developmental expansion of the information content of any given germline sequence is possible through the use of alternative RNA splicing pathways. As shown in Figure 3, antibody heavy chain genes have distinct coding regions for two relatively small, alternative carboxy-terminal tails. One of these coding regions (T_S) encodes a peptide sequence associated with the secreted form of the antibody molecule that carries out the effector functions of humoral immunity. The second of these coding regions (T_M) encodes a membrane tail which inserts into the lipid bilayer of the cell membrane and renders the antibody molecule a cell-surface receptor for triggering subsequent steps of differentiation upon interaction with a foreign molecular pattern (antigen). Though both exons are arrayed sequentially with other heavy chain exons, two alternative modes of RNA splicing ensure that only one of the two tail-exons is included in a mature message RNA (Kehry et al., 1980; Early et al., 1980; Rogers et al., 1980).

The exon-intron arrangement is complicated even further. One gene's intron may in part code for a different, structurally unrelated protein. Specifically, it has recently been shown that a protein involved in the processing of cytochrome *b* mRNA in yeast ("mRNA maturase") actually resides within the intron of the gene it helps regulate (La-zowska et al., 1980).

The hypothesis of evolutionary exon shuffling readily permits us to understand how extremely sophisticated gene systems can arise through relatively few genetic events. Therefore, under the appropriate environmental conditions, such systems also can arise in relatively short periods of evolutionary time. Let us now consider an analogous but higher order of organization for eukaryotic information that is characteristic of virtually all eukaryotic systems that have been studied to date, the multigene family.

MULTIGENE FAMILIES

Many eukaryotic genes are found in groups or families of multiple, related copies. These multigene families are made up of information units that (1) are homologous in structure, (2) overlap in function, and (3) are generally tandemly linked (Figure 4) (for review, see Hood et al., 1975). Multigene families range in size from the few gene copies of the hemoglobin families to thousands of copies of ribosomal genes and satellite DNAs. The gene copies within a family can be virtually identical or can vary markedly from one another. Multigene families are evolutionarily dynamic entities that provide special opportunities for processing, expression, expanding, and evolving information. Certain general properties of multigene families have important evolutionary implications.

1. *Variation in gene copy number.* Multigene families often vary in size from one species to another. For example, the clawed toad *Xenopus laevis* has 24,000 copies of 5S ribosomal genes, whereas its counterpart *Xenopus borealis* has 6,000 copies of the same information (Brown et al., 1971). Presumably the expansion and contraction of multigene families is promoted through homologous but unequal crossing-over (Figure 5) and can occur relatively rapidly. Variation in copy number on this scale has important implications as to the variation between copies and will be discussed shortly.

Chromosome

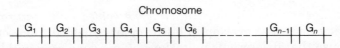

FIGURE 4. Model of a multigene family. (From Hood et al., 1975.)

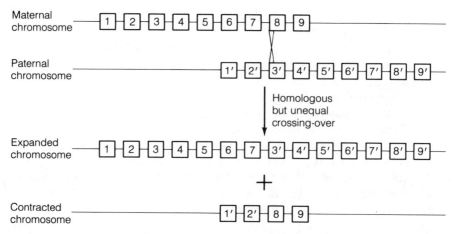

FIGURE 5. A model of homologous but unequal crossing-over. Homologous crossing-over events can be promoted by the mispairing of homologous genes between chromosomes carrying closely linked, homologous multigene families. This process yields one chromosome with an increased number of genes and a second with a decreased number of genes. (From Hood et al., 1975.)

2. *Heterogeneity*. A single gene can generally encode only a single, chemically homogeneous gene product. In contrast, a multigene family can encode a heterogeneous collection of closely related protein species. For example, the hemoglobins synthesized by humans are encoded by two multigene families (Figure 6A), the α family and the β family, whose individual genes are expressed in a developmentally regulated fashion (Figure 6B) (Bernards et al., 1979; Efstratiadis et al., 1980; Weatherall and Clegg, 1979; Proudfoot et al., 1980). Accordingly, distinct globins are expressed at the embryonic ($\zeta_2\epsilon_2$, $\alpha_2\epsilon_2$), the fetal ($\alpha_2\gamma_2$), and the adult ($\alpha_2\delta_2$, $\alpha_2\beta_2$) stages of development. Interestingly, the developmental order of β-like gene expression directly reflects the β family genomic order (5' to 3'). Also, different globins have distinct oxygen-binding properties and are regulated by various effector molecules in quite different fashions. Thus, microheterogeneity has evolved which is physiologically useful and is expressed in a precisely regulated fashion. Certainly the most extreme example of heterogeneity in the gene families analyzed to date is the diversity seen among the antibody genes, which provides the basis for the enormous range of specificities in the immune system.

3. *Multigene families shield their gene copies from natural selection*. Gene copies that overlap in function are shielded from selection, in the sense that deleterious mutations of one gene copy will be buffered if other gene copies can assume at least in part the mutated

170

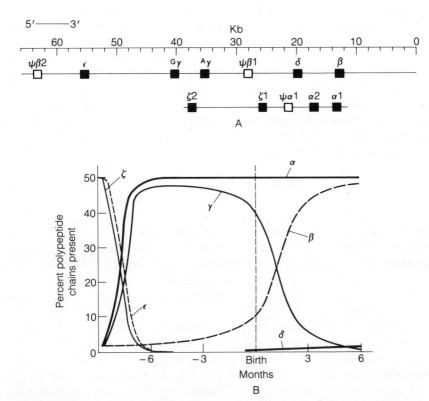

FIGURE 6. A. Models of the beta (β) and alpha (α) globin gene families in man. Individual genes (boxes) denote separate genes and do not indicate exon-intron configurations. Open boxes represent pseudogenes (ψ). The ϵ and ζ genes are embryonic; the γ genes are fetal; the α, β, and δ genes are adult. The size of these families is indicated in kilobases (kb). (Courtesy of T. Maniatis.) B. Time course of concentrations of the respective polypeptides.

gene's function. For example, hundreds of virtually identical histone genes are present in most higher organisms. The effect on the organism from the loss of function of one histone gene copy through mutation would be negligible. Thus, redundant gene copies would be expected to diverge spontaneously over evolutionary time. Besides promoting this divergence, shielding has at least two other interesting consequences. First, pseudogenes arise and appear to be a characteristic feature of multigene families (see Figure 6A). A pseudogene is one whose sequence has been altered so that it no longer is functionally expressed at the polypeptide level. Pseudogenes arise through mutations generating stop codons within exons, changes in the RNA pro-

171

cessing signals, or sequence interruptions which lead to changes in the reading frame. Indeed, 25 percent of known globin genes appear to fall in this category. The frequency of pseudogenes underscores the fact that individual members of a gene family may be under reduced selective restraint. Presumably pseudogene sequences will eventually randomize through mutation unless they have selectable functions (e.g., regulation) within the gene family (Li et al., 1981). Second, and most important, selection will tend to operate on the function of the multigene family taken as a whole rather than operating on the function of the individual gene copies that comprise that family. Thus, in one sense, the unit of selection becomes the multigene family rather than a single Mendelian gene.

4. *Coincidental evolution.* Homologous multigene families in distinct evolutionary lines can evolve in a coincidental fashion. Coincidental evolution denotes the tendency for genes within the same gene family to evolve in a similar or coincidental fashion (Figure 7). Coincidental evolution may arise either by homologous but unequal crossing-over or some type of gene conversion (Figure 8). Thus when genes are organized into tandemly linked families, there are mechanisms (Figure 8) which often insure coincidental evolution and, accordingly, limit variation.

The inherent tendency of gene copies in a multigene family to evolve coincidentally opposes their reciprocal tendency to diverge, due to their shielding from natural selection. As has been noted previously, some gene families containing hundreds of members, such as the histones, are highly uniform in structure while others, especially the antibody genes, are remarkably heterogeneous. Explanations for differences in the heterogeneity may relate to the structural characteristics of particular gene families which might increase or decrease unequal crossing-over or gene correction and the propensity for coin-

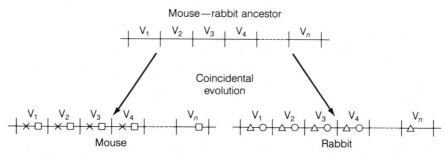

FIGURE 7. A diagram of coincidental evolution during the divergence of the mouse and rabbit evolutionary lines. ×, □, ○ and △ represent coincidental changes in the respective evolutionary lines. (From Hood et al., 1975.)

172

A. Linear transmission

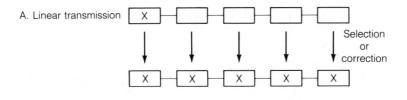

B. Transmission by
gene duplication

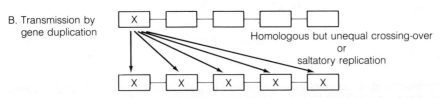

FIGURE 8. Two models of coincidental evolution. A. Linear transmission implies that all genes have been altered in a similar manner. B. Transmission by gene duplication, as in homologous but unequal crossing-over, indicates that particular gene copies have been expanded while others have been lost. (From Hood et al., 1975.)

cidental evolution. Alternatively, they may reflect structural and/or quantitative constraints at the gene product level. Copy homogeneity in gene families can be directly selected for when many functional gene copies are needed to express large quantities of product or when homogeneity of the gene product is critical for effective function or control. Ultimately the degree of homogeneity must reflect the balance between the constraints on variation imposed by the physical nature of the family and the information content necessary for that family to function.

5. *Information expansion by combinatorial and mutational mechanisms.* Multigene families can have informational and/or quantitative functions. The high copy number of ribosomal RNA genes, for example, insures the sufficient quantity of a very important, virtually homogeneous product. On the other hand, informational families like the globins produce heterogeneous products that allow an organism to fine tune its phenotype to various physiological signals (see Figure 6A, B). The antibody gene families code for specific antibody molecules against an amazingly broad spectrum of antigens. These families are predominantly informational systems that greatly expand their infor-

173

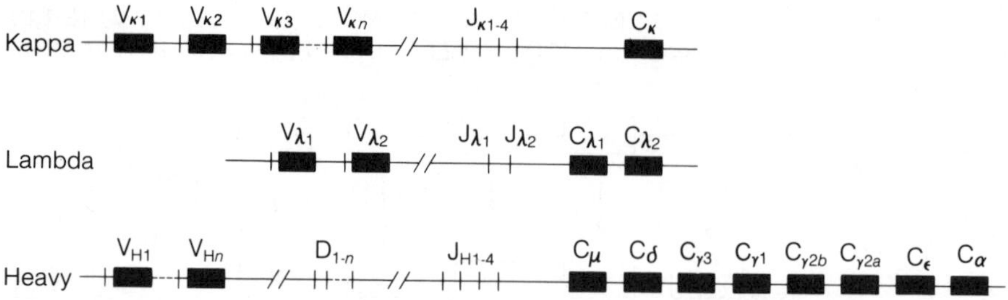

FIGURE 9. Organization of the three antibody gene families of the mouse. Exon-intron organization of heavy chain constant region genes is not shown. V, D, J, and C denote variable, diversity, joining, and constant genes, respectively. (From Early and Hood, 1981.)

mation content by DNA rearrangements and somatic mutations during development.

Antibodies are composed of two polypeptides, light and heavy chains, which in turn are divided into two distinct regions, an amino-terminal variable (V) and a carboxy-terminal constant (C) region. These polypeptides fold into four to five discrete domains which carry out distinct functions—pattern recognition by the variable domains and effector functions by the constant domains. The antibody molecules are encoded by three multigene families, two for light chains λ and κ, and a third for heavy chains (Figure 9). Light chains are

TABLE 1. Estimated numbers of antibody gene segments in the mouse and the diversity they can express via combinatorial mechanisms operating at the DNA and protein levels (see text).

1. GERMLINE		
	Kappa	$\sim$250 V_κ
		4 J_κ
	Heavy	$\sim$250 V_H
		$\sim$10 D
		4 J_H
	Lambda	2–3 V_λ
		2–3 J_λ
2. COMBINATORIAL JOINING		
	Kappa	250 $V_\kappa \times$ 4 J_κ = 1000 κ genes
	Heavy	250 $V_H \times$ 10 D $\times$ 4 J_H = 10,000 H genes
3. COMBINATORIAL ASSOCIATION		
	1000 $\kappa \times$ 10,000 H = 10^7 antibody molecules	

174

encoded by three separate gene elements, variable (V_L), joining (J_L), and constant (C_L). Heavy-chain gene families include a fourth segment, diversity (D). Table 1 gives our current estimates as to the number of germline gene segments within each of the three antibody gene families of mice.

The initial expression of antibody genes requires DNA rearrangement events that bring into contiguous apposition the elements of V_L and V_H genes with the subsequent loss of any intervening DNA sequence (Figure 10) (Brack et al., 1978; Seidman et al., 1978; Davis et al., 1980a). In light chains V_L and J_L gene elements are joined, while in the heavy chains, the V_H, D, and J_H elements are joined together. These rearrangements are presumed to occur during the development of particular antibody-producing (B) cells. Table 1 shows that unrestricted combinatorial joining of the gene elements within the antibody gene families would be equivalent to having 1,000 V_κ and 10,000 V_H germline genes in one organism. In the heavy-chain family, the joined V_H gene can be juxtaposed with different C_H segments through a developmentally regulated DNA rearrangement called class switching (Figure 10). In this way, identical antigen specificity can be associated with the various effector functions encoded by the different C_H sequences. The combinatorial association of the light and heavy

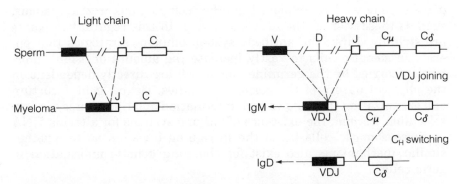

FIGURE 10. Two types of DNA rearrangements which occur during the differentiation of antibody-producing cells. In light and heavy chain genes V-J or V-D-J joining juxtaposes the gene segments encoding the V_L and V_H genes. Subsequently in heavy chain genes class, or C_H, switching may occur. The class switch leads to the expression of different immunoglobulin classes. Sperm indicates DNA undifferentiated with regard to antibody function, whereas myeloma denotes DNA in which one or more DNA rearrangements have occurred.

175

polypeptide chains is necessary to generate functional antibodies. Thus, the combinatorial pairing of light and heavy chains, at the protein level, also permits an enormous expansion of the information potential of antibody gene families and, presumably, the functional range of antigen-binding specificities. Indeed, from Table 1 we can see that 10^7 distinct antibody molecules could be generated through combinatorial mechanisms operating at both the nucleic acid and the protein level.

The mechanism that joins together the distinct gene elements has another, more subtle consequence for antibody diversity. The site of joining is flexible and may be at many different points in the junctional sequences—thereby leading to hybrid codons and junctional regions of varying size (Weigert et al., 1980; Sakano et al., 1981). Junctional diversity occurs in an area of the light and heavy chains that has a particularly important contribution to the antigen-binding site. A final, less well-understood form of antibody diversification is a mutational mechanism that appears to operate only during a narrow time span of B-cell differentiation and is probably activated by antigen stimulation (Gearhart et al., 1981; Crews et al., 1981). It causes the equivalent of point mutations in the rearranged variable gene elements and correlates with heavy-chain class switching. This mutational mechanism expands enormously the variability and hence the functional information of antibody genes and their corresponding polypeptide products.

Although presently demonstrated only in antibody gene families, these combinatorial and mutational mechanisms for information expansion may well be employed in other complex eukaryotic systems, such as those seen in the nervous system. In this regard, two points are striking. First, the antibody system employs strategies that operate in somatic cells to vastly increase the amount of useful information carried by the germline, but which are directly dependent on the physical nature of the germline families. Hence, gene structure and organization that allows such information amplification is directly selectable. Second, the existence of enzyme systems for altering DNA in these somatic cells poses the intriguing question as to whether similar mechanisms also exist for changing genetic information in germ cells.

6. *Multigene families are a new unit of evolution.* The emergence of informationally complex multigene families may have paralleled that of metazoan organisms (Figure 11). Indeed, these multigene families may reflect the enormous increase of information required at the metazoan cell surface for cellular interactions, transmembrane signaling, and cellular migrations. Once generated, these families presented the organism with a new evolutionary unit—the multigene

176

family itself. A multigene family may be duplicated in part or *in toto*. Duplicate gene families could then diverge just as duplicate genes do and come to encode new and complex aspects of eukaryotic phenotype. Since the duplication of a multigene family would include the attendant control mechanisms as well as structural genes, very sophisticated multigene families could arise in relatively short periods of evolutionary time, without the need to recreate the exhaustive and meticulous evolutionary process that was used in initially shaping the gene family. Thus, in time, a superfamily composed of many multigene families, each devoted to encoding distinct aspects of a eukaryotic phenotype, could emerge (Figure 11). This attractive hypothesis has recently been supported by the intriguing observation that genes encoding the transplantation antigens of mammals appear to be homologous to those of the antibody gene families (Strominger et al., 1980; Steinmetz et al., 1981). Thus, two complex mutigene families encoding various aspects of phenotype relating to development of the immune response appear to have a common evolutionary heritage. With the duplication of trans-active regulatory strategies, functional interaction and informational feedback schemes also can be maintained between disparate families. Again, these mechanisms permit very sophisticated gene families to evolve through a limited number of genetic events and possibly within a very short evolutionary time.

DYNAMIC GENOME—MOBILE GENETIC ELEMENTS

Perhaps the most difficult group of genetic phenomena to incorporate into the current evolutionary synthesis is represented by a staggering array of *quasi-stable genetic elements*. These are sequence units that are able to rapidly alter the genetic structure and information of an organism through rearrangement and/or expansion of genetic information, often in physiologic response to that organism's environment. Particularly interesting are mobile forms that are able to transfer genetic information between organisms, and even between species. Also, their frequent capacity for independent replication gives many of these elements a semiautonomous nature and makes them the purest example of selfish DNA: that is, DNA sequences within a cell that experience selective forces distinct from those experienced by the cell itself. Since the fitness of selfish DNA is to some degree independent of the fitness of the cell, selfish DNA is capable of independent evolution. It can be argued that all DNA possesses some selfishness. However, elements capable of interorganismic transmission are certainly less constrained by the selective needs of a particular host.

177

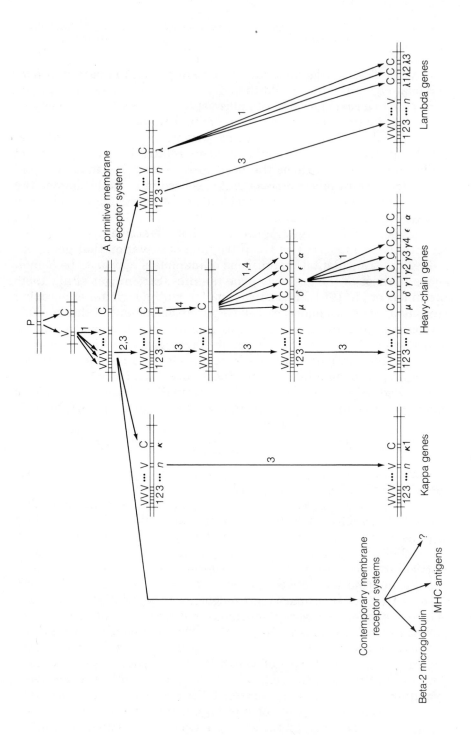

◄ **FIGURE 11.** A hypothetical scheme for the evolution of certain informational multigene families. The order of gene duplication events is unknown. A number of genetic mechanisms seem to be employed in the evolution of these families, as indicated by numbers adjacent to arrows. These are (1) discrete gene duplication, (2) gene duplication by polyploidization or chromosomal translocation, (3) contiguous gene duplication, and (4) coincidental evolution of multiple genes. Mechanisms 1 and 4 may be identical. (Adapted from Gally and Edelman, 1972.)

The mobile quasi-stable systems range in size and complexity from the tiny viroids of plants and insertion (IS) sequences of bacteria to sophisticated viruses. Their effects upon the host organism range from altering the expression of single genes to subverting the entire genome to their own purposes. Although most of the well-studied examples of mobile quasi-stable elements are found in prokaryotes, analogous systems are known or suspected in eukaryotes as well.

The best known and most widely studied group of mobile genetic elements are the transposable elements of prokaryotes (for review, see Kleckner, 1977; Bukhari et al., 1977). They are relatively short segments of DNA with direct or inverted repeats at either end. They are characterized by their participation in "illegitimate recombination" events (independent of the cell's normal recombination machinery), which may involve the duplication and translocation of their genetic information. Some are highly site specific in their transposition, while others insert randomly throughout the host's DNA. Deletion occurs independently of transposition. Similar elements that are near one another can transpose as one unit and duplicate genetic information in the DNA sequence between them, thus generating larger and more complex assemblies of mobile elements. Therefore, transposable elements represent a spectrum of forms of increasing complexity. Within this spectrum, the simpler types occur both individually and as building blocks for more complex elements, and so on to even more complex units (Figure 12).

The simplest transposable elements in bacteria are insertion sequences (IS) about 1,000 base pairs (1 kb) long with direct repeats of about 150 base pairs at either end (Figure 12). Many different types of IS are known in *E. coli,* strains of which may harbor 6 to 10 copies of an IS element. A duplication event that inserts an IS element into or adjacent to a gene can positively or negatively attenuate or totally block the expression of the gene.

179

When a phenotypically recognizable gene becomes associated with a transposable element, the unit is called a transposon (Tn) (Figure 12). Transposons carry a wide variety of genes. Most often these genes are nonessential for the normal function of the cell, but they may play a critical role under conditions of stress. The best examples, and cause of much current medical concern, are transposons that carry genes which confer resistance to antibiotics.

Transposons can be quite simple. However, by fortuitously capturing genes involved in their own replication and expression, more complex entities can arise with greater selfish characteristics. For example, element Tn117 encodes an erythromycin-resistance gene that is under repressor control of another gene specified by the same transposon. This repressor also seems to regulate transposition. Thus, exposure to erythromycin not only derepresses the resistance gene that protects the host organism but also facilitates duplication of the element.

Plasmids are autonomously replicating molecules (10^1–10^2 kb) that exist as nonintegrated or extrachromosomal bodies within the host cell (Figure 12). Conjugative plasmids constitute still higher levels of organization. The first of these to be widely studied was the fertility element (F) of *E. coli*. This element is able to initiate conjugation, a

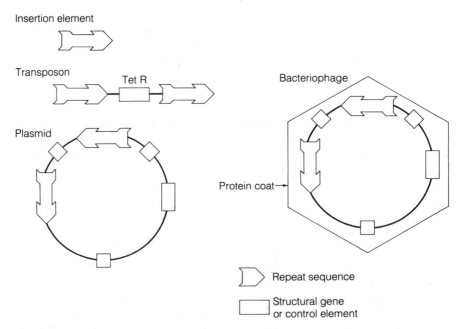

FIGURE 12. Four classes of prokaryote mobile genetic elements. Note the overlapping structural organizations.

180

process that directly transfers plasmic and chromosomal information of an F^+ cell into a recipient F^- cell. More promiscuous conjugative plasmids can even move between different species. Nonconjugative plasmids can be transferred to recipient cells during conjugation mediated by conjugative plasmids. Plasmids serve as vehicles for the cell-cell transmission of IS and transposons (Figure 12). Transposition from plasmid to host DNA can result in the stable transfer of phenotypes between the chromosomes of different organisms. Therefore, environmental conditions which promote transposition in one cell can directly facilitate the rapid acquisition of any transposition-linked trait by a very large cell population with little or no regard for that trait's selective value within the population.

The transmission of plasmids is a replication process in which the donor cell retains the same information transmitted to the recipient. Like transposons, plasmids tend to accumulate information that directly affects their own reproduction and fitness as well as the fitness of the host cells. Also, examples of both plasmids and transposons seem to have evolved through this cumulative process into still more complex units known as bacteriophages. Some members of this group are clearly plasmids (PI) or transposons (mu) that have acquired information necessary for a phage-type packaging, replication, and infection, while retaining the ability to behave in a plasmid or transposon fashion within the cell. These alternative modes of existence are determined by environmental conditions.

Many advanced phage like lambda are actually a mosaic of distinct chromosomal segments that display various modes of plasmid- and transposon-like replication, maintenance, and control. Nonlytic phage or those that integrate into the host genome also are able to provide phenotypic advantages to the cell (primarily in the form of resistance to further infection). Most important, phages represent extremely powerful vectors of genetic information between widely divergent cells.

Besides duplication of sequence information and disruption of resident gene activity, transposition is frequently associated with structural changes of the genomic material itself. Primarily, these are seen as deletions, duplications, and inversions at or near the site of transposition and may or may not include the mobile element (for review, see Nevers and Saedler, 1977).

In contrast to the above, some heritable recombination events are site specific and are directly involved in gene regulation. Inversion of a region of bacteriophage mu called the G loop appears to correlate with the production of infectious particles. In *Salmonella*, two genes (H-1 and H-2) code for the flagellar protein, flagellin. Phase variation

181

occurs when expression of one gene is replaced by expression of the other. Alternate expression of the H-1 and H-2 genes is controlled by an inversion event that changes the association between the H-2 gene and its promoter. In one orientation, H-2 is expressed along with a linked gene that encodes a repressor of H-1 expression. In the other orientation, only H-1 is expressed (Zieg et al., 1978). Interesting sequence homologies between this 1,000 bp element and Tn3 and bacteriophage mu suggest that phase variation might represent the evolutionary integration of a transposon-like selfish element into the physiologic economy of its host (Simon et al., 1980). A similar evolutionary relationship between transposable elements and many other prokaryote and eukaryote quasi-stable systems, including antibody rearrangements, has been suggested.

Although their overall importance is less well-understood, phenomena analogous to those just described for prokaryotes are displayed by eukaryotes as well as complex and poorly understood elaborations on these themes. Moreover, such phenomena appear quite common. Therefore, we believe quasi-stable elements also will play a significant evolutionary role in eukaryotic genomes.

The simplest quasi-stable eukaryotic systems are similar to bacterial insertion sequences and are exemplified by several dispersed repeated gene families in *Drosophila* (Copia, 412, etc.) (Finnegan et al., 1977; Potter et al., 1979). These elements are able to transpose with no clear pattern of distribution. They are maintained at about 25–35 copies per genome, with some variation for family type, and among individual cells, populations, and species. They are short (1 to 5 kb) with terminally redundant repeats (0.2 to 0.5 kb). As is the case with IS elements, they are frequently correlated with large chromosomal disruptions, such as deletions and inversions. Likewise, the elements may remain in place and initiate further local chromosomal disruptions. The control of transposition of these elements is not understood.

Although these elements generate no obvious protein product, members of these families are highly represented in the cytoplasmic pool of poly(A)$^+$ RNA. Consequently, some investigators have suggested a regulatory function for them. As a class, they make up perhaps 1 percent of the *Drosophila* genome. Similar DNA sequences and quasi-stable genetic elements are seen in yeast (Cameron et al., 1979) and appear to be present in many other eukaryotes. These and similar phenomena in eukaryotes are referred to specifically as mobile elements.

Eukaryotic mobile elements also exhibit a wide range of complexities analogous to prokaryote elements. For example, a series of mutations in the *white* locus (eye color) of *Drosophila* seem to be mediated by a transposon-like element derived from Copia (Rasmuson et al.,

1974). However, their nature and diversity do not always suggest the simple evolutionary connection obvious between many prokaryote mobile elements. Eukaryotes seem, in fact, to employ an even more diverse repertoire of mobile strategies.

The first phenomena to be suspected of a quasi-stable nature in any type of organism were the controlling elements of maize first studied by Rhoades and McClintock and others more than thirty years ago (for review, see Nevers and Saedler, 1977). Their apparent movement about the maize genome affected the regulation and expression of a wide variety of genetic markers. Furthermore, these changes were inherited. There are two separate component elements of maize controlling elements: (1) a receptor that inhibits gene expression and (2) a regulator that controls, transactively, the excision and transposition of itself, and the receptor. These maize-controlling elements can exist independently or linked together into an "autonomous" unit that can perform both functions. Note that the regulator has the ability to affect its own copy number in a heritable way.

In contrast to the apparently chaotic genetic changes associated with certain eukaryotic mobile elements, other quasi-stable genetic changes may be precisely controlled for gene regulation. For example, *Saccharomyces cerevisiae* (yeast) haploid cells each express one of two distinct mating types. Clonal derivatives of one type can switch to express the alternate form (for review, see Strathern and Herskowitz, 1979). Switching occurs in a cyclical manner and is under control of many genes. Mating type switch is the result of the physical replacement of one *cassette* of sequence information at the single mating-type expression locus with a new cassette from either of two alternate silent loci (Hicks et al., 1979). The silent master cassettes are not affected by mating type switching. The new mating type is then heritable.

Trypanosomes exhibit a similar cassette system for serotype determinants (Pays et al., 1981). Offspring of a single clone are able to express any one of an array of variant protein antigens by replacing the previously expressed sequence with a copy of a different one. This serotype is then inherited as a quasi-stable trait of that lineage. The original copy remains intact. The ability to vary the protein is important to the trypanosome's ability to escape the immune system of its host. The significance of all these examples is that the eukaryote genome can possess heritable traits that affect the nature of its own variation.

All examples of eukaryote quasi-stable genetic changes discussed so far produce changes that are, presumably, only vertically transmittable. The eukaryotic genome is incredibly complex, and the

183

metazoan nature of most studied eukaryotes makes horizontal transmission of information between germlines very difficult to study.
Certainly, plasmid analogs are difficult to find, except perhaps in
yeast. However, animal and plant viruses are abundant and could be
quite significant in the horizontal transmission, both within and between organisms of information that could then be inherited. Integration of viral information within eukaryote genomes is well documented. In fact, 0.1 percent of the murine genome is thought to be
related to the sequence of retroviruses, a class of RNA tumor viruses.
Related viral sequences have been found in most vertebrate genomes
studied and are closely identified with highly conserved repeat elements that resemble those of transposons (for review, see Bishop,
1978). Presumably, the repeat elements facilitate the integration of
the viral sequence into the host genome and perhaps even transposition. Many human ailments are thought to result from the activation
of information from these and other types of viruses that is normally
present but latent in the genome.

Two other examples of horizontal transmission are particularly
intriguing, because they imply exchange of genetic information over
an incredibly wide phylogenetic range. *Agrobacterium tumefaciens* is
a gram-negative soil bacterium that is the agent of crown gall neopla-
sia in many dicotyledonous plants. Oncogenic varieties of *A. tumefaciens* possess a large plasmid (Ti) that is able to conjugatively transfer
to the cells of infected plants. Much of the information is then integrated into the host cell genome where it is ready to direct the synthesis of bacteria-related products (Chilton et al., 1976; Zambryski et
al., 1980). Another class of genetic disease normally associated with
plants involves the transfer of tiny agents called viroids. These are
small (1 to 10 kb), naked single-stranded RNAs that can somehow
subvert cellular functions upon infection. Although they have been
recognized in plants for some time, recently certain animal diseases
also have been linked to viroid-like agents. Very little is known about
their biology or origin, however. An interesting hypothesis concerning
their nature is that one organism's viroid may be another's discarded
RNA. It remains to be seen whether or not viroids or bacterial plasmids can transfer heritable information into eukaryote genomes. Ultimately, however, there is no reason to believe eukaryote genomes
are less prone to violation than prokaryote genomes.

The existence and nature of quasi-stable genetic changes have
three particularly important theoretical implications. First, mobile
elements can attain the status of a parasite of the genome. This is
particularly so if originally, by its physical nature, the mobile element
mimics necessary genetic elements and becomes a competitive substrate for the enzymatic and replicative mechanisms of the host. If
transposition of an element is frequent enough, a population can be

184

parasitized with a genetic element that is irrelevant or even deleterious to the host. Also, heritable mutations within an element that affects its transposability can bring to bear unique, intragenic selective pressures and initiate competition between various transposable elements with little regard to the phenotype of the host. Of course, as with most symbiotic relationships, the parasite-host interaction will tend toward a mutualistic relationship over evolutionary time, as any benefit that the element can provide its host will reflect upon its own fitness. Thus, transposable elements are expected to evolve roles in the economy of the host genome even if they also have the capacity to evolve toward their own selfish ends. Second, transmissible elements can be seen as mobile extensions of the genome that maintain direct genetic continuity between divergent organisms. By lateral transmission, they can promote the rapid spread of selectively advantageous (e.g., antibiotic resistance) or even deleterious traits throughout populations and between species. Third, these types of recombination can generate novel genic and phenotypic arrangements with a single genetic event. These new arrangements can affect regulatory mechanisms for existing traits or even introduce entire new gene products.

Taken together, these phenomena describe a remarkably dynamic genome, one that can respond to its environment and affect its own variation. Genetic heritage must be seen as a compromise between the various selfish activities of the individual components and the selective needs of the whole.

HIERARCHICAL ORGANIZATION OF GENE EXPRESSION

The genome is dynamic but not chaotic. It is an interactive and fluid informational system best represented by hierarchical relationships. This hierarchical organization of genetic systems results in large sets of information being under common and thereby coordinate regulatory control. Therefore, it is possible for even one or very few genetic loci to control the expression of large numbers of other genetic loci. This is best exemplified by the *bithorax* complex. The *bithorax* complex is a cluster of at least 12 genetic loci that control the developmental fate of the majority of the body segments of *Drosophila melanogaster* (Lewis, 1981). *Drosophila* is a dipteran insect, characterized by having two wings on the second or mesothoracic segment and a pair of halteres on the third or metathoracic segment. The halteres are homologous to the pair of wings on the metathoracic segment on most other insects. Flies homozygous for bx^3 and *pbx,* mutant alleles at two loci in the *bithorax* complex, have four wings and no halteres (Lewis, 1963). Thus,

185

two genetic changes in the *bithorax* complex are sufficient to effect an enormous phenotypic change, presumably through the suppression of "haltere genes" and the activation of a large number of genetic loci involved in the construction of wings. Similar transformation of other segments through genetic changes in the *bithorax* complex are also seen, resulting, for example, in eight-legged flies (Lewis, 1981).

The hierarchical organization of gene expression during development has important implications for evolutionary theory. First, it makes possible major phenotypic changes that result from relatively few genetic changes, and thus rapid evolution of morphology can readily occur, provided selective conditions are favorable or neutral. It is probable that during the evolution of the Insecta, genetic changes in regulatory genes, perhaps only very few in number, created two-winged forms from the primitive four-winged state, and thus gave rise to the dipterans. It is certainly attractive to view even large-scale phyletic experiments like expansion or loss of segmentation in this context.

Second, although the new pair of wings in homozygous $bx^3 pbx$ flies are smaller than normal (Lewis, 1951, 1963), they are remarkably faithful copies of the regular wings [$bx^3 pbx$ plus mutation at a third locus, *abx*, renders the new pairs of wings virtually identical to the normal pair (Lewis, 1981)]. This leads us to believe that the second pair of wings are not the result of the activation of a cryptic set of "wing genes" specific for the metathoracic segment. Such cryptic genes should have been lost some time during the long evolutionary history of dipterans. Instead, we believe that the same set of "wing genes" are activated in a coordinated fashion in both the meso- and metathoracic segments during the development of the mutant flies. Thus the potential for the change from cryptic to manifest can be maintained for long periods of time if the genes involved are expressed elsewhere in the organism, and therefore are maintained by selection.

This potential is not limited to simple duplication or deletion of expression during development. If information that was under regulatory control of one hierarchy also is expressed within the context of other overlapping or even unrelated hierarchies, the genetic heritage of the one hierarchical order can then remain even after it is no longer coordinately expressed. Bizarre occurrences of ancestral forms are probably the best examples of this model. For instance, the enamel structure of reptilian-like teeth can still be induced in chickens (Kollar and Fisher, 1980).

Cryptic maintenance of phenotypic potential can only be one side of the coin. Changes in regulatory schemes that alter the timing of expression or even the nature of association between individual genes can result in entirely new, large-scale phenotypic experiments. The coordinate nature of the expression of such systems will tend to mitigate (though certainly not eliminate) the negative pleiotropic effects experienced upon such changes. Goldschmidt's "hopeful monster" may not be so foreign after all.

EVOLUTION AND THE NEW GENETICS

The advent of recombinant DNA technologies has led to a profound increase in our knowledge about prokaryotic and eukaryotic gene organization—indeed, it has given birth to the "new genetics," which has expanded enormously our view of the Mendelian gene and its organizational environment. This view has been very exciting. Instead of the simplicity and order many expected, an enormously complex array of structural systems and hierarchies has been found, suggesting novel mechanisms for altering genetic information. Assuming that patterns of evolution reflect the modes of genomic variation and therefore the structural and functional organization of genetic information, the perspective provided by the "new genetics" should impose specific constraints and potentialities upon evolutionary theory. Four features of the "new genetics" have particularly important implications for our views on evolution.

1. *Genomic information is organized in a hierarchical manner whose range extends from pieces of genes to the coordination of diverse multigene families and batteries of genes.* First, genes are composed of exons that often encode the individual functional domains of multidomain proteins. Combinations of domains generate sophisticated protein molecules that exhibit synergistic cooperation of functions among the domains. Second, individual genes are frequently assembled into multigene families whose members may be virtually identical or highly diverse in character. Members of the multigene family may, in turn, be simultaneously expressed, as happens to identical multigene families such as the 5S genes, or individually expressed, as happens with informational multigene families such as the antibodies. Clearly the multigene family is a primary organizational unit for control and regulation. Finally, collections or batteries of individual genes, as well as diverse multigene families, may be functionally coordinated in a hierarchical manner to generate complex patterns of development such as that seen in the *bithorax* system. Thus, units of evolutionary information must be seen as components and orders of overlapping structural and functional organizations.

187

2. *The organizational features of the genome provide it with the potential for dynamic flexibility in organization.* Prokaryotic and eukaryotic genomes have mobile elements that can move around virtually at will. The movement of these elements can change dramatically the patterns of gene expression. Their dynamic nature predicts novel methods for intra- and intergenomic transmission and expansion of information. Somatically, antibody gene element rearrangements during differentiation lead to the expression of the B-cell program for development. Both of these systems contain enzymes which can alter the structure of the genome. Thus, organisms have the enzymatic machinery to change the organization of their genomes. Almost certainly, organisms will have evolved additional, as of yet unstudied, mechanisms for changing genome organization. To what extent these elaborations can operate on germline, as opposed to somatic, DNAs is a fascinating but unanswered question.

3. *These hierarchical organizations and the dynamic properties of the genome have led to a variety of mechanisms for creating rapid and extensive variation of phenotype.* These mechanisms, once again, can operate at many levels, extending from the single subcomponents of genes to their complex hierarchical organizations. First, exons can be duplicated to create new exons with the potential for diversification and the assumption of new functions. In addition, exons can be combined together in different linear combinations to create novel genes for new multifunctional proteins generating unique combinations of synergistic functions. Second, multigene families may be duplicated in part or in their entirety to create new raw material for evolution. These duplications include the cis-attendant control mechanisms as well as the structural genes—thus very complex new gene families can be created by a single genetic event. Third, the intricate program of hierarchical gene organization and/or expression such as that seen for the *bithorax* complex can be altered—thus potentially leading to strikingly new developmental features (e.g., the placement of a haltere where wings were previously located). Thus, extremely complex organizational changes, both functional and structural, can arise from few genetic events. While the evolutionary significance of this latter class of phenomena remains at present formally unproven, it may well be that large-scale systematic and phyletic variations, such as the evolutionarily rapid increase in human brain size, or even the Cambrian explosion, will be readily understandable within this context. Also, the role of speciation and its pattern within evolution will undoubtedly be considered within this context of dramatic, often coordinated phenotypic change.

188

4. *Selection may operate at many different levels of gene organization.* Selfish DNA sequence elements exist that experience selective constraints distinct from those of their host genomes. Selection may operate on the unique advantages conferred by a certain exon and its corresponding domain. It may view as a selective unit the integrated functions of a multi-exon gene. Selection may operate on the overall expression of an entire multigene family, or it may respond to the dramatic changes in the regulatory patterns of complex developmental programs such as that of the *Drosophila bithorax* complex. Therefore, as our knowledge of how genetic information is maintained and transmitted grows, our concept of the evolutionary unit must broaden to include relational strategies that extend from the smallest identifiable genetic element to entire organisms.

Geneticists in the last five years have developed a strikingly dynamic picture of the eukaryotic genome. Given these emerging views, new theories of evolution must clearly specify the context in which terms like selection, competition, randomness, and fixation have evolutionary meaning. Obviously, molecular genetics will not provide all of the answers for understanding evolution—rather, it must be viewed as an integral aspect of a larger synthesis that will include the rich efforts of paleontology, taxonomy, and classical genetics; for in the last analysis, knowing the organism is essential.

CHAPTER 11

AUTONOMY IN EVOLUTION

John H. Campbell

This volume presents a cross section of contemporary evolutionary thought. The topics covered are quite diverse. To appreciate what they mean, it seems appropriate to borrow the evolutionists' paradigm of explaining the present by examining its historical past. Since this is the centennial of Charles Darwin's death, let us revisit the evolutionary theory of two prior Darwinian observances, the centennial and the fiftieth anniversary of the *Origin of Species*. Fortunately, Ernst Mayr, an exponent of the "synthetic" theory of Neodarwinism, has recorded the flavor of evolutionary theory as it existed 23 and 73 years ago.

When we reread the volumes published in 1909, on the occasion of the 50th anniversary of the *Origin of Species,* we realize how little agreement there was at that time among evolutionists. The change since then has been startling. Symposia and conferences were held all over the world in 1959 in honor of the Darwin centennial, and were attended by all the leading students of evolution. If we read the volumes resulting from these meetings at Cold Spring Harbor, Chicago, Philadelphia, London, Göttingen, Singapore, and Melbourne, we are almost startled at the complete unanimity in the interpretation of evolution presented by the participants. Nothing could show more clearly how internally consistent and firmly established the synthetic theory is. The few dissenters, the few who still operate with Lamarckian and finalistic concepts, display such colossal ignorance of the principles of genetics and of the entire modern literature that it would be a waste of time to refute them. The essentials of the modern theory are to such an extent consistent with the facts of genetics, systematics, and paleontology that one can hardly question their correctness. (Mayr, 1963)

The uniformity of outlook and sense of achievement in 1959 contrasts markedly with the multiplicity of views and sense of anticipation in today's volume. Obviously, 1959 marked a culmination; but what was it, and what has rejuvenated evolutionary study since then?

190

I believe that the achievement was the overwhelming acceptance of a philosophically simple cause-and-effect relationship in the evolutionary process, one that might be called Newtonian. Newtonian dynamics relies on distinguishing a mechanical system from its surroundings. A spatial boundary can be drawn around any Newtonian system to separate it categorically from its environment. Dynamic change of the system can be accounted for by two components of information. One is the current state of the system, namely the masses and velocities of its parts. The other is the external forces impinging upon these masses from the surroundings. It is the forces originating outside the system that cause the system to change.

Neodarwinism extends this framework of causality to evolution. Evolution is a process of change *in* a biological system, but *due* to forces from without. According to a traditional Neodarwinian view, the thing that is evolving is the gene pool. Its status is adequately described by the numerical frequencies of each gene allele that it contains. Evolution is defined as the change in this composition. It is caused by selective forces from the environment. Selection forces the species to evolve to fit the characteristics of the niche in the environment. Without environmental selection, the gene pool would remain constant aside from trivial or episodic stochastic changes (Founder principle), and evolution would cease. Neodarwinism represents the Newtonian mechanism right down to the algebraic paradigms by which it is described.

The triumph of a complete, mechanistic, objective, reductionistic, and mathematically precise explanation for evolution is in finally banishing vitalism from the process. Neodarwinism does so by denying the behavior of the biological system any causal role in its evolution. The gene pool does not evolve in the active sense, it "gets evolved" by the environment. The distinction between the gene pool and the environment is absolute and crucial because it coincides with the distinction between cause and effect. Current exuberance comes from a shift in interest from ideal theory to the actual process. Instead of pursuing how simply evolution could occur *theoretically,* these pages address the complexity characteristic of the way real species actually do evolve in nature. In this spirit, Stebbins' opening chapter calls for broadening evolutionary explanation from exact formulations, as in the physical sciences, to *modal* theories that admit variations in behavior of biological systems arising from their detailed individual structures.

The most significant question for evolutionary realists is whether populations are in fact constrained by the fundamental Neodarwinian dichotomy between the environment as the agency that causes and directs evolution, and the gene pool as the separate entity that passively becomes evolved. Might not the species generate its own evolutionary forces from within to augment those imposed from without? Chapter after chapter of this book suggests that, in one way or another, biological systems do participate actively in their own evolution. Biological organization at all levels, from the population down to the individual gene, can cause evolutionary changes.

In Chapter 7, Bush discusses how organization within the gene pool can cause speciation. Of course, the ideal Neodarwinian view is that speciation is caused by the environment when it impedes the flow of genes between two parts of a population. A river or mountain range divides a population and initiates the separate evolution of the two parts. Through independent adaptation of their slightly different environments, the two sister isolates gradually diverge to the levels of species and beyond. In contrast, sympatric speciation occurs when a species buds off a daughter species without intervention by the environment. Internal organization in the gene pool itself regulates gene flow and partitions off a subcomponent of the population as a reproductively isolated species. Obviously sympatric speciation will not occur in an indefinitely large, panmictic, Mendelian, ideal gene pool. The population must have internal structure or activities that lie outside of the concerns and mathematics of Neodarwinism. The potential importance of such structure of the gene pool is evident from the issue of macroevolution. The fossil record suggests to some paleontologists that species arise saltationally and subsequently persist or perish in competition with sibling species. New morphology develops mainly in the quick former phase, so it is significant if events there depend upon (or are directed by) endogenous activity of the species itself.

Spiess (1982) describes activities of individual organisms that might drive their evolution. Simplistically, the environment is the selective agent and the organisms are passively selected. However, fruit flies are not only recipients of selection but perpetrators as well. They select members of the opposite sex to become parents. How do we draw the boundary here between the selected-upon evolving unit and the selecting environment, so that the individual organisms can lie on both sides at the same time? Do we allow it the characteristics of a Möbius strip that has two discrete sides locally but only one side globally? If so, must we worry about the unexpected properties for which these complex geometrical constructions are notorious? Maybe instead we should draw around each individual a separate environment of which the rest of the population is a major part. This, however, allows the genes of a species to code directly for their own environments, and there is no place for this function of genes in Neodarwinism. Also, in some species mating (and hence "fitness") depends upon a social peck order. A main determinant of status, including reproductive inclination among monkeys, is the animal itself, especially its own internally formed perception of its status. Here, an individual animal would have to be a dominant part of its own environment. This is an unhappy situation for a Newtonian type of interpretation.

An alternative is to recognize that when a species becomes as complex as *Drosophila*—or man—it generates endogenous evolutionary forces from within, independent of the environment. This is an almost inescapable empirical conclusion. Even Darwin appreciated the existence and importance of sexual selection. However, it is anathema for the Neodarwinists who shrank its significance down to a blemish that "should not" occur. Mayr (1963) states, "Even selection with its almost unbelievable efficiency and sensitivity has a weakness in its armour [sexual selection]." But is this internal capacity for a species to generate its own evolutionary forces really a "weakness"? A living vital being that is part of the species itself would seem to contribute strength as a companion to a blind mechanical external environment for directing evolutionary change.

Several other endogenous activities that can drive evolution are mentioned in this volume, but with less emphasis than they receive elsewhere. One is mutation as a pressure for driving evolutionary change. It has been argued that point mutations become fixed in genes mainly as a consequence of internal mutation rates and not of selection (Jukes, 1980). In fact, selection can slow down the rate of DNA base-substitution during evolution (Wilson et al., 1977). Students of this role of mutation call it non-Darwinian evolution and assert that quantitatively it is the dominating process in evolutionary change (King and Jukes, 1969). Also the description of the "new genetics" with the genes being surprisingly lively touches upon the equally lively issue of selfish gene evolution. Crick and Orgel (1980) among others (e.g., Doolittle and Sapienza, 1980) have suggested that sequences of DNA can behave as parasites of the nucleus. Half of our DNA may be with us only because it autonomously replicates and evolves selfishly to enhance its own perpetuation. There is no reason to believe that DNA has to be devoid of function to enjoy this capacity: valuable genes should be equally capable of promoting their perpetuation.

Thus, the activities of biological structures at every level of organization appear capable of operating causally in the evolutionary process.

These activities breach the Neodarwinian defense against vitalism. This is no longer disturbing, because today we have developed a stronger foundation for objective biological explanation: molecular mechanism. The philosophy of molecular biology permits any form of activity, no matter how goal-directed or endogenous, as long as it is traceable to the accepted chemical behavior of macromolecules or their aggregates. If one can understand the molecular basis for a structure or a biological property emanating from that structure, the demands of molecular mechanism are satisfied. Of course, we still cannot connect the whim of a courted *Drosophila,* or the self-doubt in a monkey, back to DNA base sequence. However, we can trace activities of simpler genetic systems to DNA structure. Any principle validated for the behavior of an individual gene should be tenable for a higher level of biological organization. Accordingly, Table 1 places the activities of a variety of genes in a hierarchy of levels of complexity.

TABLE 1. Categories of structurally dynamic genes.

LEVEL 0: CLASSICAL GENES   **References**
Mendelian genes mutate only randomly and rarely.

LEVEL I: PROFANE GENES
Genes have special target sequences for gene-processing enzymes to operate upon.

A. altered sporadically

insertion sequences (transposition)	bacteria	(1)
mating type genes (cassette switching)	yeast	(2)
serotype genes (serotype variation)	trypanosomes	(3)
numerous (genotroph formation)	flax, tobacco	(4)
dissociation element Ds, a_1, a_I^{m-1}	corn	(5)
hemoglobin genes (correction)	human	(6)

B. altered in a controlled programmed manner

rRNA multigene family (amplification)	frog	(7)
tissue specific genes (amplification)	mammals	(8)
palindromic sequences (transposition)	mammals	(9)

LEVEL II: SELF-DYNAMIC GENES
Genes code for their specific gene-processing enzymes.

retroviruses (back transcription)	viruses	(10)
transposon Tn3 (transposition)	bacteria	(11)
transposon Tn554 (excision)	bacteria	(12)
inversion sequences (inversion)	bacteria, phages	(13)
controlling elements *Ac, Dt,* and *Spm*	corn	(5)

LEVEL III: CONTINGENTLY DYNAMIC GENES
Genes sense their environment and change structure in response to detected conditions.

transposon Tn917 (transposition)	bacteria	(14)
rRNA multigene family (dosage compensation)	*Drosophila*	(15)
rRNA multigene family (magnification)	*Drosophila*	(16)
genes for lysogeny (insertion)	phage lambda	(17)

194

Level O. This basal category is for the classical genes of Neodarwinism. They are inactive (in the present sense of being nonautonomous). The only way that they may change in structure is through rare, inevitable, spontaneous, and random mutations. Otherwise, they are passively copied by the cell and passively transmitted to daughter cells exactly as they were acquired from the ancestors.

TABLE 1. (*Continued*)

LEVEL IV: AUTOMODULATING GENES
Genes change their future responsiveness to stimuli when stimulated.

antibody genes (gene splicing and mutation)	vertebrates	(18)
serotype genes (serotype shift) (?)	*Paramecium*	(19)
hormone receptor genes (up and down modulation) (?)	mammals	(18)

LEVEL V: EXPERIENTIAL GENES
Genes transmit specific modifications induced in somatic phenotype to descendants.

Presumed cell surface receptor genes mediating the observed carryover for multiple generations and through male and female lines of the following acquired abnormalities.

diabetes induced by diabetogenic drugs	rats, mice, rabbits	(20)
thyroid imbalance induced by thyroidectomy or neonatal exposure to thyroxin or drugs	rats	(21)
behavioral and physiological abnormalities induced in neonatal animals by opiates	rats	(22)
disruption of pupal diapause by exposing larvae to LSD	insect	(23)
immune tolerance from injecting foreign cells into neonatal animals	mice	(24)

LEVEL VI: ANTICIPATORY/COGNITIVE GENES
Genes acquire changes in anticipation of their usefulness.

No examples identified

References

1. Bukhari et al., 1977.
2. Hicks et al., 1979.
3. Williams et al., 1979.
4. Cullis, 1977
5. McClintock, 1956; Coe and Neuffer, 1977.
6. Slightom et al., 1980.
7. Brown and Dawid, 1968.
8. Bishop and Phillips, 1978.
9. Strom et al., 1978.
10. Temin, 1974.
11. Gill et al., 1979.
12. Phillips and Novick, 1979.
13. Simon et al., 1980.
14. Tomich et al., 1978.
15. Procunier and Tartoff, 1978.
16. Ritossa, 1976.
17. Heffernan et al., 1978.
18. see Text
19. Capdeville, 1979.
20. Goldner and Spergel, 1972; Spergel et al., 1975.
21. Bakke et al., 1975.
22. Friedler, 1974; Sonderegger et al., 1979.
23. Vuillaume and Berkaloff, 1974.
24. Gorczynski and Steele, 1980, 1981.

Level I. If classical genes are analogous to sacred inviolable messages from the distant past, level I genes are profane. Organisms are allowed to modify deliberately the structure of these genes and are equipped with special gene-processing enzymes to do so. Profane genes are structurally violable because they have special target sites, generally short oligonucleotide tracts of particular base sequence, for gene-processing enzymes to recognize and operate upon. The insertion sequences described in Chapter 10 are good examples. They can be transposed to new sites along the chromosome because they are flanked at either end by a repeated nucleotide sequence that attracts transposase enzymes made by the cell (Bukhari et al., 1977). In moving to a new location they can stably alter the expression of their own information (Hicks et al., 1979) or that of genes nearby (Coe and Neuffer, 1977; Saedler et al., 1974). Table 1 illustrates that there are various sorts of profane genes. Some are modified by the cell in an occasional sporadic fashion. Others depend upon gene-processing enzymes that are under physiological control. These are active only at appropriate times. For example, the genes for ribosomal RNA in the frog form a multigene family of about 500 identical gene copies. In the oöcyte, but no other cell type, replicase enzymes expand this family to about a million gene copies (Brown and Dawid, 1968). This profoundly affects their level of expression.

Level II. Self-dynamic genes have target sites for gene-processing enzymes: in addition, they code for the enzymes that act upon them. Many bacterial transposons are self-dynamic. These genes code for useful traits, such as drug resistance, and have flanking sequences that allow them to transpose. They also contain the structural gene for their specific transposase enzyme.

Level III. A *contingently dynamic* gene specifies that it will change its structure in response to a particular external condition. Such a gene includes a coding sequence for a sensory apparatus to detect the relevant condition and to trigger activity on that basis. The transposon Tn 917 of *Streptococcus faecalis,* coding for resistance to the antibiotic erythromycin is an example (Tomich et al., 1978). The transposon contains a resistance gene, a transposase gene, flanking target sequences, and a repressor gene. The repressor controls transcription of the resistance gene, allowing expression only in the presence of erythromycin. This conforms to a widespread strategy among bacterial genes of inducibility by relevant environmental conditions. As a further development, the repressor also controls the expression of the transposase gene. Thus, in effect Tn 917 specifies that it will actively transpose only under conditions which contribute to fitness.

The behavior of another transposon, *clm,* for chloramphenicol resistance is significant here (Meyer and Iida, 1979). The resistance gene within this transposon is flanked by special internal repeat sequences. In the presence of the drug, the resistance gene within the internal repeats is reiterated into a multigene family. The enzymological basis for this change is not known, but the example indicates that the changes in gene structure are not limited to shifting the position of genes. They may be structural alterations wholly within the boundaries of the gene itself. As a point of interest, a number of drug resistance transposons are "amplifiable" (e.g., Schmitt et al., 1979; Schlöffl and Puhler, 1979).

Level IV. Contingently dynamic genes are programmed to change in structure in response to self-determined environmental or phenotypic conditions. For *automodulation*, the changes in the gene modify the detector apparatus. Hence, an external condition causes the gene to alter its future sensitivity to that condition.

The genes for antibodies reach this level of control over their information content. An antibody gene is generated by the young lymphocyte by splicing gene segments together as described in Chapter 10 (Early et al., 1980). From the recombinant genes the cell synthesizes a small number of antibody molecules with hydrophobic carboxy-terminal ends and plants them on the outer membrane as cell surface receptors for antigens. When these receptors contact the complementary antigen, a signal is relayed from the membrane back to the nucleus. It activates gene-processing enzymes to catalyze at least two, and perhaps more, structural alterations in the antibody genes and thereby change the receptors. One alteration is the stereotyped substitution of the old C-segment specifying the carboxy-terminus of the antibody molecule for a new one (Davis et al., 1980*b*). The other is the induction of scattered base-substitutions into the region of the gene that determines the antigen specificity of the antibody molecule (Gearhart et al., 1981). The rate of the expression of the gene is also markedly increased, although it is not known whether this involves a structural alteration in the antibody gene itself. After these DNA manipulations, the receptor characteristics of the gene product are changed. It is possible that antibody genes go through several cycles of automodulation as diagrammed in Figure 1 during an antibody response. Eventually, the changes prevent the antibodies from operating as cell surface receptors. This probably correlates with (or constitutes) an important event in the differentiation of the lymphocyte into its succeeding cell type, the plasma cell.

Automodulation completes a feedback loop between genotype and

197

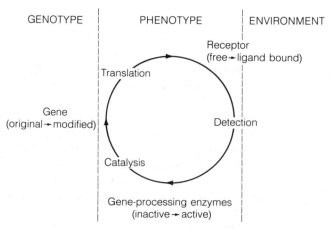

FIGURE 1. Automodulation of gene structure. The arrows trace the cycle of information among the genotype, phenotype, and environment.

phenotype. Presumably the loop could exhibit any characteristic properties of feedback, depending upon the physiological relationships among the genes' components—negative feedback, positive feedback, feedforward, oscillations, and so forth. It seems doubtful that this potential is exclusive to the antibody genes. Mammalian cells have many other cell surface receptors that are postulated to be encoded by profane genes (Hood et al., 1978) and have physiological properties suggestive of automodulation. Exposing animals at critical stages of development to abnormal levels of a variety of hormones or to drugs that mimic hormones will trigger permanent alterations in the animals' response to those hormones in the future. The basis for these changes remains speculative because the molecular structure of the genes involved have not been examined yet. I have suggested that these teratological abnormalities result from insults to automodulatory mechanisms (Campbell, 1982). Many neuroendocrine receptor genes would have the capacity to adjust their levels of expression to the hormone levels of the particular individual. The number of receptors is understood to be a major determinant of the degree of a cell's sensitivity to stimulation by a hormone (Kolata, 1977). At an early stage in ontogeny, the receptors detect the circulating levels of hormones and adjust their genes so that the number of receptor molecules expressed on the cell surface is appropriate to the genotype of the animal. This neonatal adjustment is then frozen in by dismantling the mechanism for further change. Also, various cell lineages may program their sensitivities to different levels as steps in their differentiation.

If this is the basis for the persistence so often observed in the effects of drugs on perinatal animals, it is probably more complex than simple automodulation of single genes. Sets of hormones (or *cybernins*, as this large system of diverse informational molecules is now designated; Guillemin, 1978) seem to interact in network fashion to regulate one another. The thyroid hormone system is one of many that the interested reader might explore as a possible example of complex automodulation (Bakke et al., 1975).

Level V. Changes in the structure of genes in somatic cells can be passed on to daughter cells produced by mitosis but not to an animal's progeny. For an automodulatory change to be inherited through the sexual cycle, it would have to be induced in the germ line. This is a possibility for a number of receptor genes. Germinal cells express a diversity of cybernin receptors on their cell surfaces for reasons that are not otherwise apparent (Kusano et al., 1977; El-Etr et al., 1980). Interestingly, there is suggestive evidence that a variety of altered phenotypes induced in an animal by hormones and drugs do carry over to the following generations, although alternative explanations (notably those involving threshold selection) have not been excluded.

Chemically induced diabetes is an example (Goldner and Spergel, 1972). The drug alloxan is toxic to the pancreatic β-cells that secrete insulin. It can cause permanent diabetes. Lower doses induce a subdiabetic glucose intolerance. Animals exposed to alloxan transmit glucose intolerance to their offspring. The abnormality has three striking hereditary characteristics suggesting that it is due to a change in actual DNA structure: It carries over as a specific defect in the progeny, it is transmitted by males as well as by females, and it persists for multiple generations without dying out (Spergel et al., 1975). Several other drugs that stimulate neuroendocrine receptor systems are reported to induce abnormalities with comparable genetic properties (see Table 1). Interestingly, such transgenerational effects have also been reported for the immune system (Gorczynski and Steele, 1980, 1981), although not without controversy (Brent et al., 1981; Steele, 1981; editorial, 1981; Smith, 1981). In addition, the literature reports many less well-characterized, induced carryover effects for which only one or two of the three unorthodox hereditary properties listed above have been examined (Steele, 1979; Denenberg and Rosenberg, 1967; Soyka and Joffe, 1980; Joffe, 1979; American Society of Anesthesiology, 1974; Fried and Charlebois, 1979; Lutwak-Mann, 1964; Barnett, 1973; Bresler et al., 1975; Fujii et al., 1980; Skolnick et al., 1980; Mampell, 1965, 1967, 1968; Kolata, 1977). I believe that the number of genes with some capacity to inscribe into the germ line the changes induced in the soma could be substantial. It is true, however, that some of the phenomena described are formally similar to those produced by Waddington (1953) and others through selection for susceptibility to an induced change that may also result from a rare combination of individually common alleles. In the absence of strictly controlled experiments on isogenic lines, therefore, these cases represent possibilities, not conclusions.

Level VI. The top level of endogenous gene activity for which we have molecular proof is for level IV, and level V is the limit for strongly suggestive evidence. Under the premise that we are only beginning to appreciate the range for self-activities that genes execute, Table 1 is provided with a further empty symbolic level. I have titled it "anticipatory/cognitive" to insinuate that this next level (or levels) may constitute significant qualitative advancements over level V (which in itself permits the outright inheritance of acquired characters). Might a gene change its structure in *anticipation* of an external condition or in anticipation of the effect which the change would make on phenotype? Nothing metaphysical is implied. *We* anticipate, and we experiment to determine the effects of possible changes before adopting them permanently. Nothing precludes the realization of this principle by a sufficiently sophisticated profane gene. The essence of its operation would be the complex analysis of information upon which the gene change is contingent. At least three aspects of mammalian phenotype can dynamically handle, probably analyze, and possibly build models out of complex data. They are (1) the nervous system, (2) the cybernin system, and (3) the network of interacting nucleic acid molecules that turn over in the cell nucleus including the oöcyte (Britten and Davidson, 1969; Davidson and Britten, 1979). Each of these systems is coupled to reproduction and theoretically could provide analyses of data for anticipatory gene alterations. We still are ignorant about these three essential systems of higher organisms, but all three command great interest among molecular geneticists. To date, the immune system is the only substantially complex mammalian trait for which we have any concrete genetic understanding. Its genetic mechanisms are unexpectedly elaborate. Most molecular geneticists believe that the more complex phenotypic traits still awaiting genetic characterization employ correspondingly more sophisticated manipulations of DNA molecules.

There can be no remaining doubt that genes are active, crafty (!) elements. They have complex behavior to modify and control their own structures. Also, it is clear that these endogenous activities dominate evolution in the prokaryotic kingdom (Reanney, 1976; Mitsuhashi, 1971; Anderson, 1968). For technical reasons, we know far less about eukaryotes, but it is reasonable to expect that more advanced forms of life have even cleverer genes.

200

The principle underlying the capacity of genes to evolve themselves is that the gene is not just a unit of information. It is a physical structure which contains information. When the information within the gene is directed to operate upon the physical structure which houses it, the gene can cause itself to evolve. This principle is not restricted to genes. It can emerge from any higher level of biological structure that exercises roles in propagating the information that it houses: chromosomes, germ-line cells, organisms, and populations [for example, chromosomes can generate endogenously the exact counterpart of selective forces upon themselves (Dunn, 1953, 1957; Hiraizumi et al., 1960; Brown, 1964).] Moreover, these forces can be controlled by the chromosome to levels parallel to those illustrated in Table 1 (see, for example, Pélisson and Picard, 1979; Privitera et al., 1979; Guyon et al., 1980; Petit et al., 1978; Nur, 1966; Zimmering et al., 1970).

To accept that endogenous biological behavior can cause evolutionary change is not to deny the fundamental complexity that it poses for the evolutionary process. A biological structure that directs how it will evolve (and hence how it will change the way that it will direct its evolution in the future) manifests a complicated form of causality. Its future is fundamentally less predictable than a Newtonian system, even with a "stochastic" factor added in. This mode of self-influence corresponds to an informational "strange loop" of the sort that Hofstadter (1979) has analyzed so penetratingly in his monument book *Gödel, Escher and Bach: an Eternal Golden Braid.* Systems of information that refer back to the structures that embody them are deceptively complex. In fact, they have properties that transcend the limits of formal logic! This is the basis for Gödel's incompleteness theorem of mathematics. Self-referent systems admit paradoxes that defy tractability in mathematical notation or the algebra of logic.

What a striking advance over Neodarwinism to appreciate self-reference throughout the process that created the ultimate in paradoxes, man himself. It is not surprising that, having become free of the hobgoblin of vitalism, the modern evolutionists are now intrigued by the specialities that set evolution apart from other lesser processes in nature instead of dwelling on its simpler features. What new conception of evolution emerges from this expanded vision?

This volume does not tell. The emphasis today is upon discovering ideas instead of codifying them into a new grand synthesis. We will have to wait for a future Darwinian anniversary, perhaps 2009, the bicentennary of his birth, to discover what the thoughts of today's evolutionists will beget.

201

BIBLIOGRAPHY

Abelson, J., 1979, RNA processing and the intervening sequence problem, *Ann. Rev. Biochem.*, 48, 1035–1069. (10)

Alberch, P., 1980, Ontogenesis and morphological diversification, *Amer. Zool.*, 20, 653–667. (5)

American Society of Anesthesiology, 1974, Ad hoc committee report on the effect of trace anesthetics on the health of operating personnel, *Anesthesiology*, 41, 321. (11)

Anderson, E. S., 1968, The ecology of the transferable drug resistance in the enterobacteria, *Ann. Rev. Microbiol.*, 22, 131–180. (11)

Anderson, P. K., 1980, Evolutionary implications of microtine behavioral systems on the ecological stage, *Biologist,* 62, 70–88. (2)

Anderson, S., A. T. Bankier, B. G. Barrell, M. H. L. de Bruijn, A. R. Coulson, J. Drouin, I. C. Eperon, D. P. Nierlich, B. A. Roe, F. Sanger, P. H. Schreier, A. J. H. Smith, R. Staden, and I. G. Young, 1981, Sequence and organization of the human mitochondrial genome, *Nature,* 290, 457–465. (3)

Anderson, W. F., M. G. Grütter, S. J. Remington, L. H. Weaver, and B. W. Matthews, 1981, Crystallographic determinations of the mode of binding of oligosaccharides to T4 bacteriophage lysozyme: implications for the mechanism of catalysis, *J. Mol. Biol.*, 147, 523–543. (8)

Anderson, W. W., 1973, Genetic divergence in body size among experimental populations of *Drosophila pseudoobscura* kept at different temperatures, *Evolution,* 27, 278–284. (1)

Andrews, P., 1974, New species of *Dryopithecus* from Kenya, *Nature,* 249, 188–190. (3)

Arnold, A. J., and K. Fristrup, 1982, The hierarchical basis for a unified theory of evolution, unpublished manuscript. (5)

Avise, J. C., R. A. Lansman, and R. O. Shade, 1979*a,* The use of restriction endonucleases to measure mitochondrial DNA sequence relatedness in natural populations, I. Population structure and evolution in the genus *Peromyscus, Genetics,* 92, 279–295. (3, 4)

Avise, J. C., C. Giblin-Davidson, J. Laern, J. C. Patton, and R. A. Lansman, 1979*b,* Mitochondrial DNA clones and matriarchal phylogeny within and among geographic populations of the pocket gopher, *Geomys pinetis, Proc. Natl. Acad. Sci. USA,* 76, 6694–6698. (4)

Avise, J. C., J. C. Patton, and C. F. Aquadro, 1980, Evolutionary genetics of birds: comparative molecular evolution in New World warblers, *J. Hered.*, 71, 303–310. (3)

Axelrod, D. I., 1981, Holocene climatic changes in relation to vegetation disjunction and speciation, *Amer. Nat.*, 117, 847–870. (2)

Ayala, F. J., 1974, Biological evolution: natural selection or random walk?, *Amer. Sci.*, 62, 692–701. (1)

203

Ayala, F. J., and J. A. Kiger, 1980, *Modern Genetics,* Benjamin/Cummings, Menlo Park, California. (4, 6)

Ayala, F. J., and J. W. Valentine, 1979, Genetic variability in the pelagic environment: a paradox? *Ecology,* 60, 24–29. (4)

Ayala, F. J., J. R. Powell, and T. Dobzhansky, 1971, Enzyme variability in the *Drosophila willistoni* group, II. Polymorphisms in continental and island populations of *Drosophila willistoni, Proc. Natl. Acad. Sci. USA,* 68, 2480–2483. (4)

Baba, M. L., L. L. Darga, M. Goodman, and J. Czelusniak, 1981, Evolution of cytochrome *c* investigated by the maximum parsimony method, *J. Mol. Evol.,* 17, 197–213. (1)

Bakke, J. L., N. L. Lawrence, J. Bennett, and S. Robinson, 1975, Endocrine syndromes produced by neonatal hyperthyroidism, or altered nutrition and effects seen in untreated progeny, in *Perinatal Thyroid Physiology and Disease,* D. A. Fisher and G. N. Barrow (eds.), Raven Press, New York, pp. 79–116. (11)

Barnett, S. A., 1973, Maternal processes in cold-adaptation in mice, *Biol. Rev.,* 48, 477–508. (11)

Basile, D. V., 1979, Hydroxyproline-induced changes in form, apical development and cell wall protein in the liverwort *Plagiochila arctica, Amer. J. Bot.,* 66, 776–783. (1)

Battaglia, E., 1964, Cytogenetics of B-chromosomes, *Caryologia,* 17, 245–264. (11)

Beckenbach, A. T., and S. Prakash, 1977, Examination of allelic variation at the hexokinase loci of *D. pseudoobscura* and *D. persimilis* by different methods, *Genetics,* 87, 743–761. (4)

Benyajati, C., A. R. Place, D. A. Powers, and W. Sofer, 1981, Alcohol dehydrogenase gene of *Drosophila melanogaster*: Relationship of intervening sequences to functional domains in the protein, *Proc. Natl. Acad. Sci. USA,* 78, 2717–2721. (8)

Berger, D., M. Berger, and J.-P. von Wartburg, 1974, Structural studies of human-liver alcohol-dehydrogenase isoenzymes, *Eur. J. Biochem.,* 50, 215–225. (8)

Bernards, R., P. F. R. Little, G. Annison, R. Williamson, and R. A. Flavell, 1979, Structure of the human G_γ-A_γ-δ-β-globin gene locus, *Proc. Natl. Acad. Sci. USA,* 76, 4827–4831. (10)

Bishop, J. M., 1978, Retroviruses, *Ann. Rev. Biochem.,* 47, 35–88. (10)

Bishop, J. O., and C. Phillips, 1978, in *Integration and Excision of DNA Molecules,* P. H. Hofschneider and P. Starling (eds.), Springer-Verlag, New York. (11)

Blair, W. F., 1974, Character displacement in frogs, *Amer. Zool.,* 14, 1119–1125. (1)

Blake, C. C. F., 1981, Exons and the structure, function and evolution of haemoglobin, *Nature,* 291, 616. (10)

Bock, W. J., 1979, A synthetic explanation of macroevolutionary change—a reductionistic approach, *Bull. Carnegie Mus. Nat. Hist.*, 13, 20–69. (5)

Borgaonkar, D. S., 1980, *Chromosomal Variation in Man,* 3rd ed., Alan R. Liss, Inc., New York. (7)

Borst, P., and L. A. Grivell, 1981, One gene's intron is another gene's exon, *Nature,* 289, 439–440. (7)

Boucot, A. J., 1978, Community evolution and rates of cladogenesis, in *Evolutionary Biology,* M. K. Hecht, W. C. Steere, and B. Wallace (eds.), vol. 11, Plenum Publishing Corp., New York, pp. 545–655. (5)

Brack, C., M. Hirama, R. Lenhard-Schuller, and S. Tonegawa, 1978, A complete immunoglobulin gene is created by somatic recombination, *Cell,* 15, 1–14. (10)

Brändén, C.-I., and H. Eklund, 1978, Coenzyme-induced conformational changes and substrate binding in liver alcohol dehydrogenase, in *Molecular Interactions and Activities in Proteins,* CIBA Foundation 60, Excerpta Medica, Amsterdam, pp. 63–80. (8)

Brändén, C.-I., H. Jörnvall, H. Eklund, and B. Furugren, 1975, Alcohol dehydrogenases, in *The Enzymes,* 3rd ed., P. D. Boyer (ed.), 11, pp. 103–190. (8)

Brent, L., 1981, Lamarck and immunity: the tables unturned, *New Sci.,* 90, 493. (11)

Brent, L., L. S. Rayfield, P. Chandler, W. Fierz, P. B. Medawar, and E. Simpson, 1981, Supposed Lamarckian inheritance of immunological tolerance, *Nature,* 290, 508–512. (11)

Bresler, D. E., G. Ellison, and S. Zamenhof, 1975, Learning defects in rats with malnourished grandmothers, *Dev. Psychobiol.,* 8, 315–323. (11)

Britten, R. J., and E. H. Davidson, 1969, Gene regulation for higher cells: a theory, *Science,* 165, 349–357. (11)

Britten, R. J., A. Cetta, and E. H. Davidson, 1978, The single-copy DNA sequence polymorphism of the sea urchin *Stronglylocentrotus purpuratus, Cell,* 15, 1175–1186. (4)

Brown, D. D., 1981, Gene expression in eukaryotes, *Science,* 211, 667–674. (7)

Brown, D. D., and I. B. Dawid, 1968, Specific gene amplification in oöcytes, *Science,* 160, 272–280. (11)

Brown, D. D., P. C. Wensink, and E. Jordan, 1971, Purification and some characteristics of 5S DNA from *Xenopus laevis, Proc. Natl. Acad. Sci. USA,* 68, 3175–3179. (10)

Brown, G. G. and M. V. Simpson, 1981, Intra- and interspecific variation of the mitochondrial genome in *Rattus norvegicus* and *Rattus rattus*: restriction enzyme analysis of variant mitochondrial DNA molecules and their evolutionary relationships, *Genetics,* 97, 125–143. (3, 4)

Brown, S. W., 1964, Automatic frequency response in evolution of male haploidy and other coccid chromosome systems, *Genetics,* 49, 797–817. (11)

Brown, W. M., 1980, Polymorphism in the mitochondrial DNA of humans as revealed by restriction endonuclease analysis, *Proc. Natl. Acad. Sci. USA,* 77, 3605–3609. (3)

205

Brown, W. M., M. George, Jr., and A. C. Wilson, 1979, Rapid evolution of animal mitochondrial DNA, *Proc. Natl. Acad. Sci. USA,* 76, 1967–1971. (3)

Bruce, E. J., and F. J. Ayala, 1979, Phylogenetic relationships between man and the apes: electrophoretic evidence, *Evolution,* 33, 1040–1056. (3)

Bukhari, A. I., J. A. Shapiro, and S. L. Adhya (eds.), 1977, *DNA Insertion Elements, Plasmids and Episomes,* Cold Spring Harbor Laboratory, Cold Spring Harbor, New York. (10, 11)

Bush, G. L., 1975, Modes of animal speciation, *Ann. Rev. Ecol. Syst.,* 6, 339–364. (7)

Bush, G. L., 1981, Stasipatric speciation and rapid evolution in animals, in *Evolution and Speciation: Essays in Honor of M. J. D. White,* W. R. Atchley and D. S. Woodruff (eds.), Cambridge University Press, New York, pp. 201–218. (7)

Bush, G. L., S. M. Case, A. C. Wilson, and J. L. Patton, 1977, Rapid speciation and chromosomal evolution in mammals, *Proc. Natl. Acad. Sci. USA,* 74, 3942–3946. (7)

Calame, K., J. Rogers, P. Early, M. Davis, D. Livant, R. Wall, and L. Hood, 1980, Mouse Cμ heavy chain immunoglobulin gene segment contains three intervening sequences separating domains, *Nature,* 284, 452–455. (10)

Calhoun, D. H., D. L. Pierson, and R. A. Jensen, 1973, The regulation of tryptophan biosynthesis in *Pseudomonas aeruginosa, Mol. Gen. Genet.,* 121, 117–132. (9)

Calos, M. P., and J. H. Miller, 1980, Transposable elements, *Cell,* 20, 579–595. (9)

Cameron, J. R., E. Y. Loh, and R. W. Davies, 1979, Evidence for transposition of dispersed repetitive DNA families in yeast, *Cell,* 16, 739–751. (10)

Campbell, J. H., 1982, Automodulation of genes as a mechanism for persisting effects induced by drugs, *Neurobehav. Toxicol.,* in press. (11)

Campbell, J. H., J. A. Lengyel, and J. Langridge, 1973, Evolution of a second gene for β-galactosidase in *Escherichia coli, Proc. Natl. Acad. Sci. USA,* 70, 1841–1845. (9)

Cantor, C. R., and P. R. Schimmel, 1980, *Biophysical Chemistry, Part I, The Conformation of Biological Macromolecules,* W. H. Freeman, San Francisco, p. 124ff. (8)

Capdeville, Y., 1979, Intergenic and interallelic exclusion in *Paramecium primaurelia*: immunological comparison between allelic and nonallelic surface antigens, *Immunogenetics,* 9, 77–95. (11)

Carlson, S. S., A. C. Wilson, and R. D. Maxson, 1978, Reply to "Do albumin clocks run on time?" by L. Radinsky, *Science,* 200, 1183–1185. (3)

Carson, H. L., 1959, Genetic conditions which promote or retard the formation of species, *Cold Spring Harbor Symp. Quant. Biol.,* 24, 87–105. (7)

Carson, H. L., 1975, The genetics of speciation at the diploid level, *Amer. Nat.,* 109, 83–92. (5)

Carson, H. L., and P. J. Bryant, 1979, Change in a secondary sexual character as evidence of incipient speciation in *Drosophila silvestri, Proc. Natl. Acad. Sci. USA,* 76, 1929–1932. (1)

Carson, H. L., and K. Y. Kaneshiro, 1976, *Drosophila* of Hawaii: systematics and ecological genetics, *Ann. Rev. Ecol. Syst.*, 7, 311–345. (7)

Castle, W. E., 1916, *Genetics and Eugenics,* Harvard University Press, Cambridge, Massachusetts. (1)

Cavalli-Sforza, L. L., and W. F. Bodmer, 1971, *The Genetics of Human Populations,* W. H. Freeman, San Francisco. (2)

Chambers, G. K., W. G. Laver, S. Campbell, and J. B. Gibson, 1981, Structural analysis of an electrophoretically cryptic alcohol dehydrogenase variant from an Australian population of *Drosophila melanogaster, Proc. Natl. Acad. Sci. USA,* 78, 3103–3107. (4)

Charlesworth, B., R. Lande, and M. Slatkin, 1982, A Neo-Darwinian commentary on macro-evolution, *Evolution*, 36, 474–498. (5)

Charlesworth, D., and B. Charlesworth, 1975, Theoretical genetics of Batesian mimicry. II. Evolution of supergenes, *J. Theor. Biol.*, 55, 305–324. (2)

Chatterjee, D. K., S. T. Kellog, S. Hamada, and A. M. Chakrabarty, 1981, Plasmid specifying total degradation of 3-chlorobenzoate by a modified *ortho* pathway, *J. Bacteriol.*, 146, 639–646. (9)

Cherry, J. H., 1977, Hormone action, in *The Molecular Biology of Plant Cells,* H. Smith (ed.), University of California Press, Berkeley and Los Angeles, pp. 329–364. (1)

Cherry, L. M., S. M. Case, and A. C. Wilson, 1978, Frog perspective on the morphological difference between humans and chimpanzees, *Science,* 200, 209–211. (3)

Chilton, M., S. K. Farrand, R. Levin, and E. W. Nester, 1976, RP4 promotion of transfer of a large agrobacterium plasmid which confers virulence, *Genetics,* 83, 609–618. (10)

Clarke, P. H., 1974, The evolution of enzymes for the utilization of novel substrates, in *Evolution in the Microbial World,* M. J. Carlile and J. J. Skenel (eds.), Cambridge University Press, Cambridge, England. (9)

Cochrane, B. J., and R. C. Richmond, 1979, Studies of esterase-6 in *Drosophila melanogaster.* II. The genetics and frequency distributions of naturally occurring variants studied by electrophoretic and heat-stability criteria, *Genetics,* 93, 461–478. (4)

Coe, E. H., and M. G. Neuffer, 1977, The genetics of corn, in *Corn and Corn Improvement,* G. F. Sprague (ed.), American Society of Agronomy, Madison, Wisconsin, pp. 111–223. (11)

Corruccini, R. S., J. E. Cronin, and R. L. Ciochon, 1979, Scaling analysis and congruence among anthropoid primate molecules, *Hum. Biol.*, 51, 167–185. (3)

Corruccini, R. S., M. Baba, M. Goodman, R. L. Ciochon, and J. E. Cronin, 1980, Non-linear macromolecular evolution and the molecular clock, *Evolution,* 34, 1216–1219. (3)

Coyne, J. A., and A. A. Felton, 1977, Genic heterogeneity at two *alcohol dehydrogenase* loci in *Drosophila pseudoobscura* and *Drosophila persimilis, Genetics,* 87, 285–304. (4)

Coyne, J. A., W. F. Eanes, J. A. M. Ramshaw, and R. K. Koehn, 1979, Electrophoretic heterogeneity of α-glycerophosphate dehydrogenase among many species of *Drosophila, Syst. Zool.*, 28, 164–175. (4)

207

Coyne, J. A., A. A. Felton, and R. C. Lewontin, 1978, Extent of genetic variation at a highly polymorphic *esterase* locus in *Drosophila pseudoobscura, Proc. Natl. Acad. Sci. USA,* 75, 5090–5093. (4)

Crawford, I. P., 1975, Gene rearrangements in the evolution of the tryptophan synthetic pathway, *Bacteriol. Rev.,* 39, 87–120. (9)

Crawford, I. P., 1980, Comparative studies on the regulation of tryptophan synthesis, *Crit. Revs. Biochem.,* 8, 175–189. (9)

Crawford, I. P., and I. C. Gunsalus, 1966, Inducibility of tryptophan synthetase in *Pseudomonas putida, Proc. Natl. Acad. Sci. USA,* 56, 717–724. (9)

Crawford, I. P., B. P. Nicholas, and C. Yanofsky, 1980, Nucleotide sequence of the *trpB* gene in *Escherichia coli* and *Salmonella typhimurium, J. Mol. Biol.,* 142, 489–502. (9)

Crews, S., J. Griffin, H. Huang, K. Calame, and L. Hood, 1981, A single V_H gene segment encodes the immune response to phosphorylcholine: somatic mutation is correlated with the class of the antibody, *Cell,* 25, 59–60. (10)

Crick, F., 1979, Split genes and RNA processing, *Science,* 204, 264–271. (7, 10)

Crow, J. F., and M. Kimura, 1970, *Introduction to Population Genetics Theory,* Harper & Row, New York. (6)

Crow, J. F., and M. Kimura, 1979, Efficiency of truncation selection, *Proc. Natl. Acad. Sci. USA,* 76, 396–399. (6)

Cullis, C. A., 1977, Molecular aspects of the environmental induction of heritable changes in flax, *Heredity,* 38, 129–154. (11)

Darnell, J. E., 1978, Implications of RNA-RNA splicing in evolution of eukaryotic cells, *Science,* 202, 1257–1260. (10)

Darwin, C., 1859, *On the origin of species by means of natural selection,* John Murray, London. (5)

Darwin, C. R., 1874, *The Descent of Man and Selection in Relation to Sex,* L. A. Burt, New York. (11)

Davidson, E. H., 1976, *Gene Activation in Early Development,* 2nd ed., Academic Press, New York. (1)

Davidson, E. H., and R. J. Britten, 1979, Regulation of gene expansion: possible role of repetitive sequences, *Science,* 204, 1052–1059. (11)

Davis, M., K. Calame, P. W. Early, D. L. Livant, R. Joho, I. L. Weissman, and L. Hood, 1980a, An immunoglobulin heavy-chain gene is formed by at least two recombinational events, *Nature,* 283, 733–739. (10)

Davis, M. M., S. K. Kim, and L. E. Hood, 1980b, DNA sequences mediating class switching in α-immunoglobulins, *Science,* 209, 1360–1365. (7, 11)

Dawkins, R., 1976, *The Selfish Gene,* Oxford University Press, New York. (5)

Denenberg, V. H., and K. M. Rosenberg, 1967, Nongenetic transmission of information, *Nature,* 216, 549–550. (11)

Dickerson, R. E., 1980, Cytochrome *c* and the evolution of energy metabolism, *Sci. Amer.,* 242(3), 136–153. (8)

Diderichsen, B., 1980, *Flu,* a metastable gene controlling surface properties of *Escherichia coli, J. Bacteriol.,* 141, 858–867. (9)

Dobzhansky, Th., 1951, *Genetics and the Origin of Species,* 3rd ed., Columbia University Press, New York. (7)

208

Dobzhansky, Th., 1970, *Genetics of the Evolutionary Process,* Columbia University Press, New York. (4)

Dobzhansky, Th., and F. J. Ayala, 1973, Temporal frequency changes of enzyme and chromosomal polymorphisms in natural populations of *Drosophila, Proc. Natl. Acad. Sci. USA,* 70, 680–683. (4)

Dobzhansky, Th., and B. Spassky, 1953, Genetics of natural populatons. XXI. Concealed variability in two sympatric species of *Drosophila, Genetics,* 38, 471–484. (4)

Dobzhansky, Th., and B. Spassky, 1963, Genetics of natural populations. XXXIV. Adaptive norm, genetic load and genetic elite in *D. pseudoobscura, Genetics,* 48, 1467–1485. (4)

Dobzhansky, Th., F. J. Ayala, G. L. Stebbins, J. W. Valentine, 1977, *Evolution,* W. H. Freeman, San Francisco. (4)

Dobzhansky, Th., H. Levene, B. Spassky, and N. Spassky, 1959, Release of genetic variability through recombination. III. *Drosophila prosaltans, Genetics,* 44, 75–92. (4)

Doolittle, W. F., 1978, Genes in pieces: were they ever together?, *Nature,* 581–582. (10)

Doolittle, W. F., and C. Sapienza, 1980, Selfish genes, the phenotype paradigm and genome evolution, *Nature,* 284, 601–603. (3, 5, 11)

Dunn, L. C., 1953, Variations in segregation ratio as causes of variations of gene frequency, *Acta Genet. et Statist. Med.,* 4, 139–147. (11)

Dunn, L. C., 1956, Analysis of a complex gene in the house mouse, *Cold Spring Harbor Symp. Quant. Biol.,* 21, 187–195. (5)

Dunn, L. C., 1957, Evidence of evolutionary forces leading to the spread of lethal genes in wild populations of house mice, *Proc. Natl. Acad. Sci. USA,* 43, 158–163. (11)

Dworschack, R. T., G. Tarr, and B. V. Plapp, 1975, Identification of the lysine residue modified during the activation by acetimidylation of horse liver alcohol dehydrogenase, *Biochemistry,* 14, 200–203. (8)

Early, P., and L. Hood, 1981, Mouse immunoglobulin genes, in *Genetic Engineering,* J. K. Setlow and A. Hollaender (eds.), vol. 3, Plenum Press, New York, pp. 157–188. (10)

Early, P., H. Huang, M. Davis, K. Calame, and L. Hood, 1980, An immunoglobulin heavy chain variable region gene is generated from three segments of DNA: V_H, D and J_H, *Cell,* 19, 981–992. (11)

Early, P., J. Rogers, M. Davis, K. Calame, M. Bond, R. Wall, and L. Hood, 1980, Two mRNAs can be produced from a single immunoglobulin μ gene by alternative RNA processing pathways, *Cell,* 20, 313–319. (10)

East, E. M., 1916, Inheritance in crosses between *Nicotiana langsdorffii* and *Nicotiana alata, Genetics,* 1, 311–333. (1)

Efstratiadis, A., J. W. Posakony, T. Maniatis, R. M. Lawn, C. O'Connell, R. A. Spritz, J. K. DeRiel, B. G. Forget, S. Weissman, J. L. Slightom, A. E. Blechl, O. Smithies, F. E. Baralle, C. C. Shoulders, and N. Proudfoot, 1980, Structure and evolution of the human β-globin gene family, *Cell,* 21, 653–668. (3, 10)

Eklund, H., and C.-I. Bränden, 1979, Structural differences between apo- and holoenzyme of horse liver alcohol dehydrogenase, *J. Biol. Chem.*, 254, 3458–3461. (8)

Eklund, H., C.-I. Bränden, and J. Jörnvall, 1976, Structural comparisons of mammalian, yeast and bacillar alcohol dehydrogenases, *J. Mol. Biol.*, 102, 61–73. (8)

Eklund, H., J. P. Samama, L. Wallén, C.-I. Bränden, Å. Åkeson, and T. A. Jones, 1981, Structure of a triclinic ternary complex of horse liver alcohol dehydrogenase at 2.9 Å resolution, *J. Mol. Biol.*, 146, 561–587. (8)

Eldredge, N., and J. Cracraft, 1980, *Phylogenetic Patterns and the Evolutionary Process,* Columbia University Press, New York. (5, 7)

Eldredge, N., and S. J. Gould, 1972, Punctuated equilibria: an alternative to phyletic gradualism, in *Models in Paleobiology,* T. J. M. Schopf (ed.), Freeman, Cooper and Company, San Francisco, pp. 82–115. (1, 2, 5)

El-Etr, M., S. Schorderet-Slatkine, and E. E. Baulieu, 1980, Reply, *Science,* 210, 929–930. (11)

Endler, J. A., 1977, *Geographic Variation, Speciation, and Clines,* Monographs in Population Biology, 10, Princeton University Press, Princeton, New Jersey. (7)

Eventoff, W., M. G. Rossman, S. S. Taylor, H.-J. Torff, H. Meyer, W. Keil, and H.-H. Kiltz, 1977, Structural adaptations of lactate dehydrogenase isozymes, *Proc. Natl. Acad. Sci. USA*, 274, 2677–2681. (8)

Ewens, W. J., R. S. Spielman, and H. Harris, 1981, Estimation of genetic variation at the DNA level from restriction endonuclease data, *Proc. Natl. Acad. Sci. USA,* 78, 3748–3750. (4)

Fedoroff, N. V., 1979, On spacers, *Cell,* 16, 697–710. (1)

Ferris, S. D., W. M. Brown, W. S. Davidson, and A. C. Wilson, 1981*a,* Extensive polymorphism in the mitochondrial DNA of apes, *Proc. Natl. Acad. Sci. USA,* in press. (3, 4)

Ferris, S. D., A. C. Wilson, and W. M. Brown, 1981*b,* Evolutionary tree for apes and humans based on cleavage maps of mitochondrial DNA, *Proc. Natl. Acad. Sci. USA,* 78, 2432–2436. (3)

Ferris, S. D., R. D. Sage, and A. C. Wilson, 1982, Mitochondrial DNA variation among house mice and the origins of inbred mice, *Genetics,* forthcoming. (4)

Filippi, G., A. Rinaldi, R. Palmarioni, E. Seravalli, and M. Siniscalco, 1977, Linkage disequilibrium for two X-linked genes in Sardinia and its bearing on the statistical mapping of the human X chromosome, *Genetics,* 86, 119–212. (2)

Finnegan, D. J., G. M. Rubin, M. W. Young, and D. S. Hogness, 1977, Repeated gene families in *Drosophila melanogaster, Cold Spring Harbor Symp. Quant. Biol.,* 42, 1053–1063. (10)

Fisher, R. A., 1930, *The genetical theory of natural selection,* Clarendon Press, Oxford, England. (1)

Fitch, W. M., 1966, An improved method of testing evolutionary homology, *J. Mol. Biol.*, 16, 9–16, (3)

210

Fitch, W. M., 1976a, Molecular evolutionary clocks, in *Molecular Evolution,* F. J. Ayala (ed.), Sinauer Associates, Sunderland, Massachusetts, pp. 160–178. (1, 3)

Fitch, W. M., 1976b, The molecular evolution of cytochrome c in eucaryotes, *J. Mol Evol.,* 8, 13–40. (9)

Fitch, W. M., and C. H. Langley, 1976, Evolutionary rates in proteins: neutral mutations and the molecular clock, in *Molecular Anthropology,* M. Goodman et al. (eds.), Plenum Press, New York, pp. 197–219. (3)

Fitch, W. M., and E. Margoliash, 1967, Construction of phylogenetic trees, *Science,* 155, 279–284. (3)

Flavell, R. B., 1981, Molecular changes in chromosomal DNA organization and origins of phenotypic variation, *Chromosomes Today,* 7, 42–54. (7)

Fletcher, T. S., 1979, Biochemical analysis of *Drosophila melanogaster Adh* alleles, Ph.D. dissertation, University of California, Davis. (4)

Ford, E. B. 1975, *Ecological Genetics,* 4th ed., John Wiley and Sons, New York. (7)

Fox, G. E., L. J. Magrum, W. E. Balch, R. S. Wolfe, and C. R. Woese, 1977, Classification of methogenic bacteria by 16S ribosomal RNA characterization, *Proc. Natl. Acad. Sci. USA,* 74, 4537–4541. (3)

Fox, G. E., E. Stackebrandt, R. B. Hespell, J. Gibson, J. Maniloff, T. A. Dyer, R. S. Wolfe, W. E. Balch, R. S. Tanner, L. J. Mangrum, L. B. Zablen, R. Blakemore, R. Gupta, L. Bonen, B. J. Lewis, D. A. Stahl, K. R. Leuhrsen, K. N. Chen, and C. R. Woese, 1980, The phylogeny of procaryotes, *Science,* 209, 457–463. (3, 9)

Fried, P. A., and A. T. Charlebois, 1979, Cannabis administered during pregnancy—1st- and 2nd-generation effects in rats, *Physiol. Psych.,* 7, 307–310. (11)

Friedler, G., 1974, Long term effects of opiates, in *Perinatal Pharmacology: Problems and Priorities,* J. Dancis and J. C. Hwang (eds.), Raven Press, New York, pp. 207–219. (11)

Fujii, T., S. Morimoto, and H. Ikeda, 1980, Hypersensitivity to lethal effects of injected calcium chloride in the third generation of rats raised from parathyroidectomized mothers, *Biomed. Res.,* 1, 432–434. (11)

Futuyma, D. J., and G. C. Mayer, 1980, Non-allopatric speciation in animals, *Syst. Zool.,* 29, 254–271. (7)

Gally, J. A., and G. M. Edelman, 1972, The genetic control of immunoglobulin synthesis, *Ann. Rev. Genet.,* 6, 1–46. (10)

Gearhart, P., N. Johnson, R. Douglas, and L. Hood, 1981, IgG antibodies to phosphorylcholine exhibit more diversity than their IgM counterparts, *Nature,* 291, 29–34. (10, 11)

Gerschitz, J., R. Rudolph, and R. Jaenicke, 1978, Refolding and reactivation of liver alcohol dehydrogenase after dissociation and denaturation in 6M guanidine hydrochloride, *Eur. J. Biochem.,* 87, 591–599. (8)

Ghiselin, M. T., 1974, A radical solution to the species problem, *Syst. Zool.,* 26, 437–438. (5)

Gifford, E. M. J., and G. E. Corson, Jr., 1971, The shoot apex in seed plants, *Bot. Rev.,* 37, 143–229. (1)

Gilbert, W., 1978, Why genes in pieces?, *Nature,* 271, 501. (10)

Gilinsky, N. L., 1981, Stabilizing species selection in the Archaeogastropoda, *Paleobiology,* 7(3), 316–331. (5)

Gill, R. E., F. Heffron, and S. Falkow, 1979, Identification of the protein encoded by the transposable element Tn3 which is required for its transposition, *Nature,* 282, 797–801. (11)

Gliddon, C., and C. Strobeck, 1975, Necessary and sufficient conditions for multiple-niche polymorphism in haploids, *Amer. Nat.,* 109, 233–235. (6)

Goldner, M. G., and G. Spergel, 1972, On the transmission of alloxan diabetes and other diabetogenic influences, *Adv. Metab. Disord.,* 6, 57–72. (11)

Goldschmidt, R. B., 1940, *The Material Basis of Evolution,* Yale University Press, New Haven, Connecticut. (5, 7)

Goodman, M. 1976*a*, Protein sequences in phylogeny, in *Molecular Evolution,* F. J. Ayala (ed.), Sinauer Associates, Sunderland, Massachusetts, pp. 141–159. (1)

Goodman, M., 1976*b*, Toward a genealogical description of the primates, in *Molecular Anthropology,* M. Goodman et al. (eds.), Plenum Press, New York, pp. 321–353. (3)

Goodman, M., 1981*a*, Decoding the pattern of protein evolution, *Progress in Biophysics and Molecular Biology,* in press. (3)

Goodman, M., 1981*b*, Globin evolution was apparently rapid in early vertebrates: a reasonable case against the rate-constancy hypothesis, *J. Mol. Evol.,* 17, 114–120. (3)

Goodman, M., G. W. Moore, and G. Matsuda, 1975, Darwinian evolution in the genealogy of haemoglobin, *Nature,* 253, 603–608. (3)

Goodman, M., R. E. Tashian, and J. H. Tashian (eds.), 1976, *Molecular Anthropology, Genes and Proteins in the Evolutionary Ascent of the Primates,* Plenum Press, New York. (3)

Gorczynski, R. M., and E. J. Steele, 1980, Inheritance of acquired immunological tolerance to foreign histocompatibility antigens in mice, *Proc. Natl. Acad. Sci. USA,* 77, 2871–2875. (11)

Gorczynski, R. M., and E. J. Steele, 1981, Simultaneous yet independent inheritance of somatically acquired tolerance to two distinct H-2 antigenic haplotype determinants in mice, *Nature,* 289, 678–681. (11)

Gottlieb, L. D., 1981, Electrophoretic evidence and plant populations, *Phytochemistry,* 7, 1–46. (4)

Gould, S. J., 1970, Dollo on Dollo's law: irreversibility and the status of evolutionary laws, *J. Hist. Biol.,* 3, 189–212. (5)

Gould, S. J., 1977, *Ontogeny and Phylogeny,* Harvard University Press, Cambridge, Massachusetts, p. 501. (5)

Gould, S. J., 1980, Is a new and general theory of evolution emerging?, *Paleobiology,* 6, 119–130. (5)

Gould, S. J., 1981, G. G. Simpson, paleontology and the modern synthesis, in *The Evolutionary Synthesis,* E. Mayr and W. Provine (eds.), Harvard University Press, Cambridge, Massachusetts, pp. 153–172.

Gould, S. J., 1982, *The uses of heresy: an introduction to R. Goldschmidt's: The Material Basis of Evolution,* Yale University Press, New Haven, Connecticut. (5)

Gould, S. J., and N. Eldredge, 1977, Punctuated equilibria: the tempo and mode of evolution reconsidered, *Paleobiology,* 3, 115–151. (5)

Grantham, R., C. Gautier, M. Gouy, and R. Mercier, 1981, Codon catalog usage is a genome strategy modulated for gene expressivity, *Nucleic Acids Res.,* 9, r43–r74. (9)

Green, P. B., 1969, Cell morphogenesis, *Ann. Rev. Plant Physiol.,* 20, 365–394. (1)

Green, P. B., R. O. Erickson, and P. A. Richmond, 1970, On the physical basis of cell morphogenesis, *Ann. New York Acad. Sci.,* 175, 712–731. (1)

Green, P. B., R. O. Erickson, and J. Buggy, 1971, Metabolic and physical control of cell elongation rate. *In vivo* studies in *Nitella, Plant Physiol.,* 47, 423–430. (1)

Greenfield, L. M., 1974, Taxonomic reassessment of two *Ramapithecus* specimens, *Folia Primatol.,* 22, 97–115. (3)

Gruber, H. E., 1974, *Darwin on Man: a Psychological Study of Scientific Creativity,* E. P. Dutton, New York. (5)

Grula, J. W., T. J. Hall, J. A. Hunt, T. D. Guigni, E. H. Davidson, and R. J. Britten, 1982, Sea urchin DNA sequence variation and reduced interspecies differences of the less variable DNA sequences, *Evolution,* 36, in press. (4)

Guillemin, R., 1978, Peptides in the brain: the new endocrinology of the neuron, *Science,* 202, 390–402. (11)

Guyon, P., M.-D. Chinton, A. Petit, and J. Temple, 1980, Agropine in "null-type" crown gall tumors. Evidence for generality of the opine concept, *Proc. Natl. Acad. Sci. USA,* 77, 2693–2697. (11)

Haldane, J. B. S., and S. D. Jayakar, 1963, Polymorphism due to selection of varying direction, *J. Genet.,* 58, 237–242. (6)

Hall, B. G., 1977, Number of mutations required to evolve a new lactase function in *Escherichia coli, J. Bacteriol.,* 129, 540–543. (9)

Hall, B. G., 1978, Regulation of newly evolved enzymes. IV. Directed evolution of the *ebg* repressor, *Genetics,* 90, 673–681. (9)

Hall, B. G., and D. L. Hartl, 1974, Regulation of newly evolved enzymes. I. Selection of a novel lactase regulated by lactose in *Escherichia coli, Genetics,* 76, 391–400. (9)

Hallam, A., 1978, How rare is phyletic gradualism and what is its evolutionary significance? Evidence from Jurassic bivalves, *Paleobiology,* 4, 16–25. (5)

Hansen, T., 1978, Larval dispersal and species longevity in Lower Tertiary gastropods, *Science,* 199, 885–887. (5)

Hansen, T. A., 1980, Influence of larval dispersal and geographic distribution on species longevity in neogastropods, *Paleobiology,* 6, 193–207. (5)

Hartl, D., and D. Dykhuizen, 1979, A selectively driven molecular clock, *Nature,* 281, 230–231. (3)

Hedrick, P., S. Jain, and L. Holden, 1978, Multilocus systems in evolution, in *Evolutionary Biology,* M. K. Hecht, W. C. Steere, and B. Wallace (eds.), vol. 11, Plenum Publishing Corp., New York, pp. 101–184. (7)

Hedrick, P. W., and M. E. Ginevan, and E. P. Ewing, 1976, Genetic polymorphism in heterogeneous environments, *Ann. Rev. Ecol. Syst.,* 7, 1–32. (6)

213

Heffernan, L., M. Benedik, and A. Campbell, 1978, Regulatory structure of the insertion region of bacteriophage lambda, *Cold Spring Harbor Symp. Quant. Biol.*, 43, 1127–1134. (11)

Henderson, E. J., and H. Zalkin, 1971, On the composition of anthranilate synthetase-anthranilate 5-phosphoribosylpyrophosphate phosphoribosyltransferase from *Salmonella tryphimurium*, *J. Biol. Chem.*, 246, 6891–6898. (9)

Hicks, J., N. Strathern, and A. J. S. Klar, 1979, Transposable mating-type genes in *Saccharomyces cerevisiae*, *Nature*, 282, 478–483. (10, 11)

Hiraizumi, Y., L. Sandler, and J. F. Crow, 1960, Meiotic drive in natural populations of *Drosophila melanogaster*. III. Population implications of the *segregation-distortion* locus, *Evolution*, 14, 433–444. (11)

Hoch, S O., C. Anagnostopoulos, and I. P. Crawford, 1969, Enzymes of the trypotophan operon of *Bacillus subtilis*, *Biochem. Biophys. Res. Comm.*, 35, 838–844. (9)

Hofstadter, D., 1979, *Gödel, Escher, Bach: an Eternal Golden Braid*, Basic Books, New York. (11)

Holmquist, R., 1978, The augmentation algorithm and molecular phylogenetic trees, *J. Mol. Evol.*, 12, 17–24. (3)

Holmquist, R., 1980, Evolutionary analysis of α and β hemoglobin genes by REH theory under the assumption of the equiprobability of genetic events, *J. Mol. Evol.*, 15, 149–159. (3)

Hood, L., J. H. Campbell, and S. C. R. Elgin, 1975, The organization, expression, and evolution of antibody genes and other multigene families, *Ann. Rev. Genet.*, 9, 305–353. (10)

Hood, L., H. V. Huang, and W. J. Dreyer, 1978, The area code hypothesis: the immune system provides clues to understanding the genetic and molecular basis of cell recognition during development, *J. Supramol. Str.*, 7, 531–559. (11)

Hull, D. L., 1980, Individuality and selection, *Ann. Rev. Ecol. Syst.*, 11, 311–332. (5)

Iino, T., and K. Kutsukake, 1980, *Trans*-acting genes of bacteriophages Pl and Mu mediate inversion of a specific DNA segment involved in flagellar phase variation in *Salmonella, Cold Spring Harbor Symp. Quant. Biol.*, 45, 11–16. (11)

Jablonski, D., 1980, Apparent versus real biotic effects of transgressions and regresssions, *Paleobiology*, 6, 397–407. (5)

Jacobs, L. L., and D. Pilbeam, 1980, Of mice and men: fossil-based divergence dates and molecular "clocks," *J. Hum. Evol.*, 9, 551–555. (3)

Jacoby, G. A., and J. A. Shapiro, 1977, Plasmids studied in *Pseudomonas aeruginosa* and other pseudomonads, in *DNA Insertion Elements, Plasmids, and Episomes*, A. I. Bukhari, J. A. Shapiro, and S. Adhya (eds.), Cold Spring Harbor Laboratory, Cold Spring Harbor, New York, pp. 639–656. (9)

Jeffreys, A. J., 1979, DNA sequence variants in the $^G\gamma$-, $^A\gamma$-, δ- and β-globin genes of man, *Cell*, 18, 1–10. (4)

214

Joffe, J. M., 1979, Influence of drug exposure of the father on perinatal outcome, *Clinics in Perinatology,* 6, 21–36. (11)

Johnson, J. L., and B. S. Francis, 1975, Taxonomy of the *Clostridia*: ribosomal RNA homologies among the species, *J. Gen. Microbiol.,* 88, 229–244. (9)

Jones, J. S., and R. F. Probert, 1980, Habitat selection maintains a deleterious allele in a heterogeneous environment, *Nature,* 287, 632–633. (6)

Jörnvall, H., H. Eklund, C.-I. Brändén, 1978, Subunit conformation of yeast alcohol dehydrogenase, *J. Biol. Chem.,* 253, 8414–8419. (8)

Jukes, T. H., 1980, Silent nucleotide substitutions and the molecular clock, *Science,* 210, 973–978. (3, 11)

Kane, J. F., W. M. Holmes, and R. A. Jensen, 1972, Metabolic interlock: the dual function of a folate pathway gene as an extraoperonic gene of tryptophan biosynthesis, *J. Biol. Chem.,* 247, 1587–1506. (9)

Kaplan, N., and C. H. Langley, 1979, A new estimate of sequence divergence of mitochondrial DNA using restriction endonuclease mapping, *J. Mol. Evol.,* 13, 295–304. (3)

Kavenoff, R., L. C. Klotz, and B. H. Zimm, 1974, On the nature of chromosome-sized DNA molecules, *Cold Spring Harbor Symp. Quant. Biol.,* 38, 1–8. (1)

Kehry, M., S. Ewald, R. Douglas, C. Sibley, W. Raschke, D. Fambrough, and L. Hood, 1980, The immunoglobulin μ chains of membrane-bound and secreted IgM molecules differ in their C-terminal segments, *Cell,* 21, 393–406. (10)

Kellogg, D. E., 1975, The role of phyletic change in the evolution of *Pseudocubus vema* (Radiolaria), *Paleobiology,* 1, 359–370. (5)

Kettlewell, B., 1973, *The Evolution of Melanism,* Clarendon, New York. (2)

Kipling, R., 1912, *Just So Stories,* Doubleday and Co., New York. (7)

Kimura, M., 1968*a,* Evolutionary rate at the molecular level, *Nature,* 217, 624–626. (3, 6)

Kimura, M., 1968*b,* Genetic variability maintained in a finite population due to mutational production of neutral and nearly neutral isoalleles, *Genet. Res.,* 11, 247–269. (3)

Kimura, M., 1974, Gene pool of higher organisms as a product of evolution, *Cold Spring Harbor Symp. Quant. Biol.,* 38, 515–524. (7)

Kimura, M., 1977, Preponderance of synonymous changes as evidence for the neutral theory of molecular evolution, *Nature,* 267, 275–276. (3)

Kimura, M., 1979, Model of effectively neutral mutations in which selective constraint is incorporated, *Proc. Natl. Acad. Sci. USA,* 76, 3440–3444. (6)

Kimura, M., 1980, A simple method for estimating evolutionary rates of base substitutions through comparative studies of nucleotide sequences, *J. Mol. Evol.,* 16, 111–120. (3)

Kimura, M., 1981*a,* Was globin evolution very rapid in its early stages?: a dubious case against the rate-constancy hypothesis, *J. Mol. Evol.,* 17, 110–113. (3)

Kimura, M., 1981*b,* Doubt about studies of globin evolution based on maximum parsimony codons and the augmentation procedure, *J. Mol. Evol.,* 17, 121–122. (3)

215

Kimura, M., 1981c, Possibility of extensive neutral evolution under stabilizing selection with special reference to non-random usage of synonymous codons, *Proc. Natl. Acad. Sci. USA,* 78, 5773–5777. (6)

Kimura, M., and J. F. Crow, 1978, Effect of overall phenotypic selection on genetic change at individual loci, *Proc. Natl. Acad. Sci. USA,* 75, 6168–6171. (6)

King, J. L., and T. H. Jukes, 1969, Non-Darwinian evolution, *Science,* 164, 788–798. (3, 11)

King, M.-C., and A. C. Wilson, 1975, Evolution at two levels in humans and chimpanzees, *Science,* 188, 107–116. (5)

Kitamura, N., H. L. Semler, P. G. Rothberg, G. R. Larsen, C. J. Adler, A. J. Dorner, E. A. Emini, R. Hanecak, J. J. Lee, S. van der Werf, C. W. Anderson, and E. Wimmer, 1981, Primary structure, gene organization and polypeptide expression of poliovirus RNA, *Nature,* 291, 547–553. (3)

Kleckner, N., 1977, Translocatable elements in procaryotes, *Cell,* 11, 11–23. (9, 10)

Koehn, R. K., and W. F. Eanes, 1978, Molecular structure and protein variation within and among populations, in *Evolutionary Biology,* M. K. Hecht, W. C. Steere, and B. Wallace (eds.), vol. 11, Plenum Publishing Corp., New York, pp. 39–100. (4)

Kolata, G. B., 1977, Hormone receptors: how are they regulated?, *Science,* 196, 747–749, 800. (11)

Kolata, G. B., 1978, Teratogens acting through males, *Science,* 207, 733. (11)

Kollar, E. J., and C. Fisher, 1980, Tooth induction in chick epithelium: expression of quiescent genes for enamel synthesis, *Science,* 207, 993–995. (10)

Koo, G. C., S. S. Wachtel, K. Krupen-Brown, L. R. Mittl, W. R. Berg, M. Genel, I. M. Rosenthal, D. S. Borgaonkar, D. A. Miller, R. Tantravahi, R. R. Schreck, B. F. Erlanger, and O. J. Miller, 1977, Mapping the locus of the H-Y gene on the human Y chromosome, *Science,* 198, 940–942. (7)

Korey, K. A., 1981, Species number, generation length, and the molecular clock, *Evolution,* 35, 139–147. (3)

Kreitman, M., 1980, Assessment of variability within electromorphs of alcohol dehydrogenase in *Drosophila melanogaster, Genetics,* 95, 467–475. (4)

Kurtén, B., 1972, *Not from the Apes,* Vintage Books, New York. (3)

Kusano, K., R. Miledi, and J. Stinnaker, 1977, Acetylcholine receptors in the oöcyte membrane, *Nature,* 270, 739–741. (11)

Lai, E. C., S. L. C. Woo, A. Dugaiczyk, and B. W. O'Malley, 1979, The ovalbumin gene: alleles created by mutations in the intervening sequences of the natural gene, *Cell,* 16, 201–212. (4)

Lande, R., 1976, Natural selection and random genetic drift in phenotypic evolution, *Evolution,* 30, 314–334. (1)

Lande, R., 1978, Evolutionary mechanisms of limb loss in tetrapods, *Evolution,* 32, 73–92. (5)

Lande, R., 1979, Effective deme size during long-term evolution estimated from rates of chromosomal rearrangement, *Evolution,* 33, 234–251. (7)

216

Lande, R., 1980, Microevolution in relation to macroevolution, *Paleobiology,* 6, 233–238. (5)

Langley, C. H., and W. M. Fitch, 1974, An examination of the constancy of the rate of molecular evolution, *J. Mol. Evol.,* 3, 161–177. (3)

Lansman, R. A., R. O. Shade, J. R. Shapira, and J. C. Avise, 1981, The use of restriction endonucleases to measure mitochondrial DNA sequence relatedness in natural populations. III. Techniques and potential applications, *J. Mol. Evol.,* 17, 214–226. (3)

Larson, A., D. B. Wake, L. R. Maxson, and R. Highton, 1981, A molecular phylogenetic perspective on the origins of morphological novelties in the salamanders of the tribe Plethodontini (Amphibia, Plethodontidae), *Evolution,* 35, 405–422. (3)

Lawn, R. M., E. F. Fritsch, R. C. Parker, G. Blake, and T. Maniatis, 1978, The isolation and characterization of linked δ- and β-globin genes from a cloned library of human DNA, *Cell,* 15, 1157–1174. (4)

Lazowska, J., C. Jacq, and P. P. Slonimski, 1980, Sequence of introns and flanking exons in wild-type and *box3* mutants of cytochrome *b* reveals an interlaced splicing protein coded by an intron, *Cell,* 22, 333–348. (10)

Le Minor, L., and R. Rohde, 1974, *Salmonella,* in *Bergey's Manual of Determinative Bacteriology,* R. E. Buchanan and N. E. Gibbons (eds.), Williams and Wilkins Co., Baltimore, pp. 298–317. (9)

Lessios, H. A., 1979, Use of Panamanian sea urchins to test the molecular clock, *Nature,* 280, 599–601. (3)

Lessios, H. A., 1981, Divergence in allopatry: molecular and morphological differentiation between sea urchins separated by the Isthmus of Panama, *Evolution,* 35, 618–634. (3)

Levene, H., 1953, Genetic equilibrium when more than one ecological niche is available, *Amer. Nat.,* 87, 331–333. (6)

Levitt, M., and C. Chothia, 1976, Structural patterns in globular proteins, *Nature,* 261, 552–558. (3, 8)

Lewin, R., 1980, Evolutionary theory under fire, *Science,* 210, 883–887. (2)

Lewin, R., 1981, How conversational are genes?, *Science,* 212, 313–315. (7)

Lewis, E. B., 1951, Pseudoallelism and gene evolution, *Cold Spring Harbor Symp. Quant. Biol.,* 16, 159–174. (10)

Lewis, E. B., 1963, Genes and developmental pathways, *Amer. Zool,* 3, 33–56. (10)

Lewis, E. B., 1981, ICN-UCLA Symposia, in press. (10)

Lewontin, R. C., 1970, The units of selection, *Ann. Rev. Ecol. Syst.,* 1, 1–18. (5)

Lewontin, R. C., 1974, *The Genetic Basis of Evolutionary Change,* Columbia University Press, New York. (6)

Li, W., T. Gojobori, and M. Nei, 1981, Pseudogenes as a paradigm of neutral evolution, *Nature,* 292, 237–239. (3, 10)

Lindsley, D. L., and E. H. Grell, 1968, Genetic variation of *Drosophila melanogaster, Carnegie Inst. of Wash.,* Pub. 627. (7)

Loukas, M., Y. Vergini, and C. B. Krimbas, 1981, The genetics of *Drosophila subobscura* populations. XVII. Further genic heterogeneity within electromorphs by urea denaturation and the effect of the increased genic variability on linkage disequilibrium studies, *Genetics,* 97, 429–441. (4)

Lowenstein, J. M., 1980, Species-specific proteins in fossils, *Naturwissenschaften,* 67, 343–346. (3)

Lowenstein, J. M., V. M. Sarich, and B. J. Richardson, 1981, Albumin systematics of the extinct mammoth and Tasmanian wolf, *Nature,* 290, 409–411. (3)

Lutwak-Mann, C., 1964, Observations on progeny of thalidomide-treated male rabbits, *Brit. Med. J.,* 1, 1090–1091. (11)

Mampell, K., 1965, Gene expression and developmental rate, *Genetica,* 36, 135–146. (11)

Mampell, K., 1967, Genetic and environmental control of melanotic tumors in *Drosophila, Genetica,* 37, 449–465. (11)

Mampell, K., 1968, Differentiation and extragenic transmission of modified gene expression, *Genetica,* 39, 553–556. (11)

Manch, J. M., and I. P. Crawford, 1981, Ordering tryptophan synthase genes of *Pseudomonas aeruginosa* by cloning in *Escherichia coli, J. Bacteriol.,* 146, 102–107. (9)

Manney, T. R., W. Duntze, N. Janosko, and J. Salazar, 1969, Genetic and biochemical studies of partially active tryptophan synthetase mutants of *Sacharomyces cerevisiae, J. Bacteriol.,* 99, 590–596. (9)

Margoliash, E., W. M. Fitch, and R. E. Dickerson, 1969, Molecular expression of evolutionary phenomena in the primary and tertiary structures of cytochrome *c,* in *Structure, Function and Evolution of Proteins,* vol. II, Brookhaven Symp. Biol. 21, Brookhaven National Laboratory, Upton, New York, pp. 259–305. (3)

Marshall, D. R., and A. H. D. Brown, 1975, The charge-state model of protein polymorphism in natural populations, *J. Mol. Evol.,* 6, 149–163. (4)

Marshall, L. G., 1977, Cladistic analysis of borhyaenoid, dasyuroid, didelphoid, and thylacinid (Marsupalia: Mammalia) affinity, *Syst. Zool.,* 26, 410–425. (3)

Matchett, W. H., and J. A. DeMoss, 1975, The subunit structure of tryptophan synthase from *Neurospora crassa, J. Biol. Chem.,* 250, 2941–2946. (9)

Matheson, A. T., W. Möller, R. Amons, and M. Yaguchi, 1980, Comparative studies on the structure of ribosomal proteins, with emphasis on the alanine-rich, acidic ribosomal, 'A' protein, in *Ribosomes: Structure, Function, and Genetics,* G. Chamblis, G. R. Craven, J. Davies, K. Davis, L. Kahan, and M. Nomura (eds.), University Park Press, Baltimore, pp. 297–332. (3)

Matthews, B. W., S. J. Remington, M. G. Grütter, and W. F. Anderson, 1981, Relation between hen egg-white lysozyme and bacteriophage T4 lysozyme: evolutionary implications, *J. Mol. Biol.,* 147, 545–558. (3, 8)

Maurer, R., and I. P. Crawford, 1971, New regulatory mutant affecting some of the tryptophan genes in *Pseudomonas putida, J. Bacteriol.,* 106, 331–338. (9)

Maxson, L. R., and R. D. Maxson, 1979, Comparative albumin and biochemical evolution in plethodontid salamanders, *Evolution,* 33, 1057–1062. (3)

218

Maxson, L. R., and A. C. Wilson, 1979, Rates of molecular and chromosome evolution in salamanders, *Evolution,* 33, 734–740. (1)

Maynard Smith, J., 1966, Sympatric speciation, *Amer. Nat.*, 100, 637–650. (7)

Mayr, E., 1942, *Systematics and the Origin of Species,* Columbia University Press, New York. (1)

Mayr, E., 1954, Change of genetic environment and evolution, in *Evolution as a Process,* J. Huxley, A. C. Hardy, and E. B. Ford (eds.), Allen and Unwin, London, pp. 157–180. (1, 7)

Mayr, E., 1963, *Animal Species and Evolution,* Harvard University Press, Cambridge, Massachusetts. (5, 7, 11)

Mayr, E., 1969, *Principles of Systematic Zoology,* McGraw-Hill Book Co., New York. (3)

Mayr, E., 1978, Evolution, *Sci. Amer.*, 239(3), 46–55. (5)

Mayr, E., 1981, Biological classification: toward a synthesis of opposing methodologies, *Science*, 214, 510–516. (3)

McClintock, B., 1956, Controlling elements and the gene, *Cold Spring Harbor Symp. Quant. Biol.*, 21, 197–216. (11)

McDonald, J. F., and F. J. Ayala, 1978, Genetic and biochemical basis of enzyme activity variation in natural populations. I. Alcohol dehydrogenase in *Drosophila melanogaster, Genetics,* 89, 371–388. (4)

McDonald, J. F., G. K. Chambers, J. David, and F. J. Ayala, 1977, Adaptive response due to changes in gene regulation: a study with *Drosophila, Proc. Natl. Acad. Sci. USA,* 74, 4562–4566. (4)

Meyer, J., and S. Iida, 1979, Amplification of chloramphenicol resistance transposons carried by phage *P1cm* in *Escherichia coli.*, *Mol. Gen. Genet.*, 176, 209–219. (11)

Milkman, R., 1970, The genetic basis of natural variation in *Drosophila, Advances in Genetics,* 15, 55–114. (11)

Milkman, R., 1973, Electrophoretic variation in *E. coli* from natural sources, *Science,* 182, 1024–1026. (6)

Milkman, R., 1978*a,* Selection differentials and selection coefficients, *Genetics,* 88, 391–403. (6)

Milkman, R., 1978*b,* The maintenance of polymorphisms by natural selection, in *Marine Organisms: Genetics, Ecology, Evolution,* J. A. Beardmore and B. Battaglia (eds.), Plenum Press, New York, pp. 3–22. (6)

Minty, A., and P. Newmark, 1980, Gene regulation: new, old and remote controls, *Nature,* 288, 210–211. (7)

Miozzari, G. F., and C. Yanofsky, 1979, Gene fusion during the evolution of the tryptophan operon in Enterobacteriaceae, *Nature,* 277, 486–489. (9)

Mitsuhashi, S. (ed.), 1971, *Transferable Drug Resistance Factor R,* University Park Press, Baltimore. (11)

Miyata, T., T. Yasunaga, and T. Nishida, 1980, Nucleotide sequence divergence and functional constraint in mRNA evolution, *Proc. Natl. Acad. Sci. USA,* 77, 7328–7332. (3)

Moore, R. C., and L. R. Laudon, 1943, Evolution and classification of Paleozoic crinoids, *Geol. Soc. Amer. Spec. Papers,* No. 46, 153 pp. (5)

Mourão, C. A., F. J. Ayala, and W. W. Anderson, 1972, Darwinian fitness and adaptedness in experimental populations of *Drosophila willistoni, Genetica,* 43, 552–574. (4)

Nakamura, K., and M. Inouye, 1980, DNA sequence of the *Serratia marcescens* lipoprotein gene, *Proc. Natl. Acad. Sci. USA,* 77, 1369–1373. (9)

Nei, M., 1979, Stochastic theory of population genetics and evolution, in *Lecture Notes in Biomathematics,* 39, C. Barigozzi (ed.), Springer-Verlag, New York, 17–47. (7)

Nei, M. and W.-H. Li, 1979, Mathematical model for studying genetic variation in terms of restriction endonucleases, *Proc. Natl. Acad. Sci. USA,* 76, 5269–5273. (3)

Nei, M., and W.-H. Li, 1980, Non-random association between electromorphs and inversion chromosomes in finite populations, *Genet. Res. Camb.,* 35, 65–83. (7)

Nei, M., and F. Tajima, 1981, DNA polymorphism detectable by restriction endonucleases, *Genetics,* 97, 145–163. (3)

Nevers, P., and H. Saedler, 1977, Translocatable genetic elements as agents of gene instability and chromosomal rearrangements, *Nature,* 268, 109–115. (10)

Nevo, E., 1978, Genetic variation in natural populations: patterns and theory, *Theor. Pop. Biol.,* 13, 121–177. (4)

Nevo, E., and H. Cleve, 1978, Genetic differentiation during speciation, *Nature,* 275, 125–126. (1)

Nichols, B. P., and C. Yanofsky, 1979, Nucleotide sequences of *trpA* of *Salmonella typhimurium* and *Escherichia coli:* an evolutionary comparison, *Proc. Natl. Acad. Sci. USA,* 76, 5244–5248. (9)

Nichols, B. P., M. Blumenberg, and C. Yanofsky, 1981, Comparison of the nucleotide sequence of *trpA* and sequences immediately beyond the *trp* operon of *Klebsiella aerogenes, Salmonella typhimurium* and *Escherichia coli, Nucleic Acids Res.,* 7, 1743–1755. (9)

Nichols, B. P., G. F. Miozzari, M. van Cleemput, G. N. Bennett, and C. Yanofsky, 1980, Nucleotide sequences of the *trpG* regions of *Escherichia coli, Shigella dynsenteriae, Salmonella typhimurium* and *Serratia marcescens, J. Mol. Biol.,* 142, 503–517. (9)

Nilsson-Ehle, H., 1909, Kreuzungssuntersuchungen an Hafer und Weizen, *Lunds Universitets Arsskrift,* n.s. 2, vol. 5, no. 2. (1)

Nobs, M. A., 1963, Experimental Studies on Species Relationships in *Ceanothus, Carnegie Inst. of Wash.,* Pub. 623, p. 94. (1)

Noller, H. F., and C. R. Woese, 1981, Secondary structure of 16S ribosomal RNA, *Science,* 212, 403–411. (3, 9)

Nur, U., 1966, Nonreplication of heterochromatic chromosomes in a mealy bug *Planococcus citra* (Coccoidea-Homoptera), *Chromosoma,* 19, 439. (12)

Nuti, M. P., A. A. Lepidi, R. K. Prakash, R. A. Schilperoort, and F. C. Cannon, 1979, Evidence for nitrogen fixation (*rif*) genes on indigenous *Rhizobium* plasmids, *Nature,* 282, 533–535. (9)

220

Ohlsson, I., B. Nordstrom, and C.-I. Brändén, 1974, Structural and functional similarities within the coenzyme binding domains of dehydrogenases, *J. Mol. Biol.*, 89, 339–354. (8)

Ohta, T., 1974, Mutational pressure as the main cause of molecular evolution and polymorphism, *Nature*, 252, 351–354. (1, 6)

Orgel, L. E., and F. H. C. Crick, 1980, Selfish DNA—the ultimate parasite, *Nature*, 284, 604–607. (3, 5, 11)

Osawa, S., and H. Hori, 1980, Molecular evolution of ribosomal components, in *Ribosomes: Structure, Function, and Genetics,* G. Chamblis, G. R. Craven, J. Davies, K. Davis, L. Kahan, and M. Nomura (eds.), University Park Press, Baltimore, pp. 333–355. (3)

Oster, G., and P. Alberch, 1982, Evolution and bifurcation of developmental programs, *Evolution*, 36, in press. (1)

Pace, B., and L. L. Campbell, 1971, Homology of ribosomal RNA of diverse bacterial species with *Escherichia coli* and *Bacillus stearothermophilus, J. Bacteriol.*, 107, 543–547. (9)

Pays, E., N. Van Meirvenne, D. Le Ray, and M. Stewart, 1981, Gene duplication and transposition linked to antigenic variation in *Trypanosoma brucei, Proc. Natl. Acad. Sci. USA,* 78, 2673–2677. (10)

Pélisson, A., and G. Picard, 1979, Non-Mendelian female sterility in *Drosophila melanogaster*—I-factor mapping on inducer chromosomes, *Genetica,* 50, 141–148. (11)

Pembrey, M. E., R. P. Perrine, W. G. Wood, and D. J. Weatherall, 1980, Sickle-β° Thalassemia in Eastern Saudi Arabia, *Amer. J. Hum. Genet.*, 32, 26–41. (2)

Perler, F., A. Efstratiadis, P. Lomedico, W. Gilbert, R. Kolodner, and J. Dodgson, 1980, The evolution of genes: the chicken preproinsulin gene, *Cell,* 20, 555–566. (3)

Petit, A., J. Tempe, A. Kerr, M. Holsters, M. Van Montague, and J. Schell, 1978, Substrate induction of conjugative activity of *Agrobacterium tumefaciens* Ti plasmids, *Nature,* 271, 570–572. (11)

Phillips, S., and R. P. Novick, 1979, Tn554—a site-specific repressor-controlled transposon in *Staphylococcus aureus, Nature,* 278, 476–478. (11)

Pipkin, S., and N. Hewitt, 1972, Variation of alcohol dehydrogenase levels in *Drosophila* species hybrids, *J. Hered.*, 63, 267. (4)

Plapp, B. V., 1970, Enhancement of the activity of horse liver alcohol dehydrogenase by modification of amino groups at the active sites, *J. Biol. Chem.*, 245, 1727–1735. (8)

Plapp, B. V., H. Eklund, and C.-I. Brändén, 1978, Crystallography of liver alcohol dehydrogenase complexed with substrates, *J. Mol. Biol.*, 122, 23–32. (8)

Post, L. E., G. D. Stycharz, M. Nomura, H. Lewis, and P. P. Dennis, 1977, Ribosomal protein gene cluster adjacent to gene for RNA polymerase subunit β in *Escherichia coli*: nucleotide sequences, *Proc. Natl. Acad. Sci. USA,* 76, 1697–1701. (9)

Potter, S. S., W. J. Brorein, P. Dunsmuir, and G. M. Rubin, 1979, Transposition of elements of the *412, copia,* and *297* dispersed repeated gene families in *Drosophila, Cell,* 17, 415–427. (10)

Powell, J. R., 1975, Protein variation in natural populations of animals, *Evol. Biol.,* 8, 79–119. (4)

Prager, E. M., A. C. Wilson, J. M. Lowenstein, and V. M. Sarich, 1980, Mammoth albumin, *Science,* 209, 287–289. (3)

Prakash, S., and R. C. Lewontin, 1968, A molecular approach to the study of genic heterozygosity in natural populations. III. Direct evidence of coadaptation in gene arrangements of *Drosophila, Proc. Natl. Acad. Sci. USA,* 59, 398–405. (7)

Privitera, G., M. Sebald, and F. Fayolle, 1979, Common regulatory mechanism of expression and conjugative ability of a tetracycline resistant plasmid in *Bacteroides fragilis, Nature,* 278, 657–659. (11)

Procunier, J. D., and K. D. Tartoff, 1978, A genetic locus having trans and contiguous cis functions that control the disproportionate replication of the ribosomal RNA genes of *Drosophila melanogaster, Genetics,* 88, 76–79. (11)

Proudfoot, N., 1980, Pseudogenes, *Nature,* 286, 840–841. (3)

Proudfoot, N. J., M. H. M. Shander, J. L. Manley, M. L. Gefter, and T. Maniatis, 1980, Structure and *in vitro* transcription of human globin genes, *Science,* 209, 1329–1336. (10)

Prout, T., 1968, Sufficient conditions for multiple-niche polymorphism, *Amer. Nat.,* 102, 493–497. (6)

Provine, W., *A Biography of Sewall Wright,* in preparation. (5)

Provine, W. B., 1971, *The Origins of Theoretical Population Genetics,* University of Chicago Press, Chicago. (1)

Ramshaw, J. A. M., J. A. Coyne, and R. C. Lewontin, 1979, The sensitivity of gel electrophoresis as a detector of genetic variation, *Genetics,* 93, 1019–1037. (4)

Rasmuson, B., M. M. Green, and B. Karlsson, 1974, Genetic instability in *Drosophila melanogaster, Mol. Gen. Genet.,* 133, 237–247. (10)

Raup, D. M., 1978, Cohort analysis of generic survivorship, *Paleobiology,* 4, 1–15. (5)

Reanney, D. C., 1976, Extrachromosomal elements as possible agents of adaptation and development, *Bacteriol. Rev.,* 40, 552–590. (11)

Regier, J. C., and F. C. Kafatos, 1981, *In vivo* kinetics of pyrrolidonecarboxylic acid formation in selected silkmoth chorion proteins, *J. Biol. Chem.,* 256, 6444–6451. (8)

Richardson, J., 1977, β-sheet topology and the relatedness of proteins, *Nature,* 268, 495–500. (8)

Ritossa, F., 1976, *Bobbed* locus, in *The Genetics and Biology of Drosophila,* M. Ashburner and E. Novitski (eds.), vol. 1b, Academic Press, London, pp. 801–846. (11)

Roberts, R. J., 1981, Restriction and modification enzymes and their recognition sequences, *Nucleic Acids Res.,* 9, r75–r96. (9)

222

Robertson, A., 1956, The effect of selection against extreme deviants based on deviation or on homozygosis, *J. Genet.*, 54, 236–248. (6)

Robertson, A., 1977, Why are mice the size they are: in *Proceedings of the International Conference on Quantitative Genetics,* E. Pollak, O. Kempthorne, and T. B. Bailey, Jr. (eds.), Iowa State University Press, Ames, Iowa. (6)

Rodriguez, R., (ed.), 1981, *Promoters: Structure and Function,* Praeger, New York. (1)

Rogers, J., P. Early, C. Carter, K. Calame, M. Bond, L. Hood, and R. Wall, 1980, Two mRNAs with different 3' ends encode membrane-bound and secreted forms of immunoglobulin μ chain, *Cell,* 20, 303–312. (10)

Rosenblum, I. M., 1981, An approach toward understanding some of the morphogenetic bases of the phylogeny of *Streptocarpus* (Gesneriaceae), Ph.D. dissertation, City University of New York, Lehman College, New York. (1)

Rosenzweig, M. L., 1978, Competitive speciation, *Biol. J. Linn. Soc.,* 10, 275–289. (7)

Rossmann, M. G., and P. Argos, 1976, Exploring structural homology of proteins, *J. Mol. Biol.,* 105, 75–95. (3)

Rossmann, M. G., and P. Argos, 1978, The taxonomy of binding sites in proteins, *Mol. Cell. Biochem.,* 21, 161–182. (3)

Rossmann, M. G., A. Liljas, C.-I. Brändén, and L. J. Banazak, 1975, Evolutionary and structural relationships among dehydrogenases, in *The Enzymes,* 3rd ed., P. D. Boyer (ed.), vol. 11, pp. 61–102. (8)

Roth, Jr., E. F., G. Schiliro, A. Russo, S. Musumeci, E. Rachmilewitz, V. Neske, and R. Nagel, 1980, Sickle cell disease in Sicily, *J. Med. Genet.,* 17, 34–38. (2)

Ruse, M., 1980, Charles Darwin and group selection, *Ann. of Sci.,* 37, 615–630. (5)

Saedler, H., J. H. Reif, S. Hu, and N. Davidson, 1974, *IS2,* a genetic element for turn-off and turn-on of gene activity in *E. coli, Mol. Gen. Genet.,* 132, 265–289. (11)

Sakano, H., Y. Kuiosawa, M. Weigert, and S. Tonegawa, 1981, Identification and nucleotide sequence of a diversity DNA segment (D) of immunoglobulin heavy-chain genes, *Nature,* 290, 562–565. (10)

Sakano, H., J. A. Rogers, K. Huppi, C. Brack, A. Traunecker, R. Maki, R. Wall, and S. Tonegawa, 1979, Domains and the hinge region of an immunoglobulin heavy chain are encoded in separate DNA fragments, *Nature,* 277, 627–634. (10)

Sampsell, B., 1977, Isolation and genetic characterization of alcohol dehydrogenase thermostability variants occurring in natural populations of *Drosophila melanogaster, Biochem. Gen.,* 15, 971–988. (4)

Sanderson, K. E., 1976, Genetic relatedness in the family Enterobacteriaceae, *Ann. Rev. Microbiol.,* 30, 327–349. (3)

Sarich, V. M., 1977, Rates, sample sizes, and the neutrality hypothesis for electrophoresis in evolutionary studies, *Nature,* 265, 24–28. (3)

Sarich, V. M., and J. E. Cronin, 1976, Molecular systematics of the primates, in *Molecular Anthropology,* M. Goodman et al. (eds.), Plenum Press, New York, pp. 141–170. (3)

Sarich, V. M., and J. E. Cronin, 1980, South American mammal molecular systematics, evolutionary clocks, and continental drift, in *Evolutionary Biology of the New World Monkeys and Continental Drift,* R. L. Ciochon and A. B. Chiarelli (eds.), Plenum Press, New York, pp. 399–421. (3)

Sarich, V. M., and A. C. Wilson, 1967a, Rates of albumin evolution in primates, *Proc. Natl. Acad. Sci. USA,* 58, 142–148. (3)

Sarich, V. M., and A. C. Wilson, 1967b, Immunological time scale for hominid evolution, *Science,* 158, 1200–1203. (3)

Satoh, C., and H. W. Mohrenweiser, 1979, Genetic heterogeneity within an electrophoretic phenotype of phosphoglucose isomerase in a Japanese population, *Ann. Hum. Genet.,* London, 42, 283–292. (4)

Saunders, J. W., 1968, *Animal Morphogenesis,* Macmillan, New York and London. (1)

Scanlan, B. E., L. R. Maxson, and W. E. Duellman, 1980, Albumin evolution in marsupial frogs (Hylidae: *Gastrotheca*), *Evolution,* 34, 222–229. (3)

Schindel, D. E., 1980, Microstratigraphic sampling and the limits of paleontologic resolution, *Paleobiology,* 6, 408–426. (5)

Schindewolf, O. H., 1950, *Grundfragen der Paläontologie,* E. Schweizerbart, Stuttgart. (5)

Schlöffl, F., and A. Puhler, 1979, Intramolecular amplification of the tetracycline resistance determinant of transposon Tn1771 in *Escherichia coli, Genet. Res.,* 33, 253–260. (11)

Schmitt, R., E. Bernhard, and R. Mattes, 1979, Characterization of Tn1721, a new transposon containing tetracycline genes capable of amplification, *Mol. Gen. Genet.,* 172, 53–65. (11)

Schopf, J. W., 1978, The evolution of the earliest cells, *Sci. Amer.,* 239(3), 110–138. (1)

Schreier, P. H., A. L. M. Bothwell, B. Mueller-Hill, and D. Baltimore, 1981, Multiple differences between the nucleic acid sequences of the IgG2a[a] and IgG2a[b] alleles in the mouse, *Proc. Natl. Acad. Sci. USA,* 78, 4495–4499. (4)

Schulz, G. E., and R. H. Schirmer, 1979, *Principles of Protein Structure,* Springer-Verlag, Berlin. (8)

Sciaky, D., A. L. Montoya, and M.-D. Chilton, 1978, Fingerprints of *Agrobacterium* Ti plasmids, *Plasmid,* 1, 238–253. (9)

Seager, R. D., 1979, Fitness interactions and genetic load in *Drosophila melanogaster,* Ph.D. dissertation, University of California, Davis. (4)

Seidman, J. G., A. Leder, M. H. Edgell, F. Polsky, S. M. Tilghman, D. C. Tiemeier, and P. Leder, 1978, Multiple related immunoglobulin variable-region genes identified by cloning and sequence analysis, *Proc. Natl. Acad. Sci. USA,* 75, 3881–3885. (10)

Selander, R. K., 1976, Genic variation in natural populations, in *Molecular Evolution,* F. J. Ayala (ed.), Sinauer Associates, Sunderland, Massachusetts, pp. 21–45. (4, 6)

Selander, R. K., and B. R. Levin, 1980, Genetic diversity and structure in *Escherichia coli* populations, *Science,* 210, 545–547. (6)

Shuto, T., 1974, Larval ecology of prosobranch gastropods and its bearing on biogeography and paleontology, *Lethaia,* 7, 239–256. (5)

Simon, M., J. Zieg, M. Silverman, G. Mandel, and R. Doolittle, 1980, Phase variation: evolution of a controlling element, *Science,* 209, 1370–1374. (9, 10, 11)

Simons, E. L., 1976, The fossil record of primate phylogeny, in *Molecular Anthropology,* M. Goodman et al. (eds.), Plenum Press, New York, pp. 35–62. (3)

Simons, E. L., 1977, *Ramapithecus, Sci. Amer.,* 236(5), 28–35. (3)

Simons, E. L., and D. R. Pilbeam, 1978, *Ramapithecus* (Hominidae, Hominoidea), in *Evolution of African Mammals*, V. J. Maglio and H. B. S. Cooke (eds.), Harvard University Press, Cambridge, Massachusetts, pp. 147–153. (3)

Simpson, G. G., 1941, The affinities of the Borhyaenidae, *Amer. Mus. Novit.,* 1118, 1–6. (3)

Simpson, G. G., 1944, *Tempo and Mode in Evolution,* Columbia University Press, New York. (5)

Simpson, G. G., 1945, The principles of classification and a classification of mammals, *Bull. Amer. Mus. Nat. Hist.,* 85, 1–350. (3)

Simpson, G. G., 1953, *The Major Features of Evolution,* Columbia University Press, New York. (5)

Simpson, G. G., 1963, The meaning of taxonomic statements, in *Classification and Human Evolution,* S. L. Washburn (ed.), Aldine, Chicago, pp. 1–31. (3)

Singh, R. S., 1979, Genetic heterogeneity within electrophoretic "alleles" and the pattern of variation among loci in *Drosophila pseudoobscura, Genetics,* 93, 997–1018. (4)

Skolnick, N. J., S. H. Ackerman, M. A. Hofer, and H. Weiner, 1980, Vertical transmission of acquired ulcer susceptibility in the rat, *Science,* 208, 1161–1163. (11)

Slaughter, C. A., M. C. Coseo, M. P. Cancro, and H. Harris, 1981, Detection of enzyme polymorphism by using monoclonal antibodies, *Proc. Natl. Acad. Sci. USA,* 78, 1124–1128. (4)

Slightom, J. L., A. E. Blechi, and O. Smithies, 1980, Human fetal $^G\gamma$- and $^A\gamma$-globin genes: complete nucleotide sequences suggest that DNA can be exchanged between these duplicated genes, *Cell,* 21, 627–638. (4, 11)

Smith, R. M., 1981, Inability of tolerant males to sire tolerant progeny, *Nature,* 292, 767–768. (11)

Sober, E., 1982, Holism, individualism and the units of selection, *Phil. of Sci. Assoc.,* in press. (5)

Sober, H. A., 1970, *Handbook of Biochemistry: Selected Data for Molecular Biology,* 2nd ed., CRC Press, West Palm Beach, Florida, pp. H83–H85. (9)

Sogin, D. C., and B. V. Plapp, 1975, Activation and inactivation of horse liver alcohol dehydrogenase with pyridoxal compounds, *J. Biol. Chem.,* 250, 205–210. (8)

Sogin, D. C., and B. V. Plapp, 1976, Inactivation of horse liver alcohol dehy-drogenase by modification of cysteine residue 174 with diazonium-1*H*-tetrazole, *Biochemistry,* 15, 1087–1093. (8)

Somero, G. N., and M. Soulé, 1974, Genetic variation in marine fishes as a test of the niche-variation hypothesis, *Nature,* 249, 670–672. (4)

Sonderegger, T., S. O'Shea and E. Zimmermann, 1979, Progeny of male rats addicted neonatally to morphine, *Proc. West. Pharmacol. Soc.,* 22, 137–139. (11)

Soyka, L. F., and J. M. Joffe, 1980, Male-mediated drug effects on offspring, in *Drug and Chemical Risks to the Fetus and New Born,* R. H. Schwarz and S. J. Yaffe (eds.), Alan R. Liss, Inc., New York, pp. 49–66. (11)

Spergel, G., F. Kahn, and M. G. Goldner, 1975, Emergence of overt diabetes in offspring of rats with latent diabetes, *Metabolism,* 24, 1311–1319. (11)

Spiess, E., 1982, Sexual selection, *Amer. Nat.,* in press. [Presidential address to The American Society of Naturalists, Iowa City, June, 1981.] (11)

Stanley, S. M., 1975, A theory of evolution above the species level, *Proc. Natl. Acad. Sci. USA,* 72, 646–650. (5)

Stanley, S. M., 1979, *Macroevolution: Pattern and Process,* W. H. Freeman, San Francisco. (1, 2, 5)

Stebbins, G. L., 1965, Some relationships between mitotic rhythm, nucleic acid synthesis and morphogenesis in higher plants, Brookhaven Symp. Biol., 18, *Genetic Control of Differentiation,* 204–221. (1)

Stebbins, G. L., 1968, Gene action, mitotic frequency and morphogenesis in higher plants, *Developmental Biology,* Suppl. I, 26th Symp. Soc. Devel. Biol., *Control Mechanisms in Developmental Processes,* Academic Press, New York, pp. 113–135. (1)

Stebbins, G. L., 1974, *Flowering Plants: Evolution Above the Species Level,* Harvard University Press, Cambridge, Massachusetts. (1)

Stebbins, G. L., 1982, Plant speciation, in *Mechanisms of Speciation,* C. Bar-igozzi (ed.), Publ. Accad. Naz. dei Lincei, Rome, in preparation. (1)

Stebbins, G. L., and F. J. Ayala, 1981, Is a new evolutionary synthesis nec-essary?, *Science,* 213, 967–971. (1, 5)

Stebbins, R. C., 1949, Speciation in salamanders of the Plethodontid genus *Ensatina, Univ. Calif. Publ. Zool.,* 48, 377–526. (1)

Steele, E. J., 1979, *Somatic Selection and Adaptive Evolution: On the Inher-itance of Acquired Characters,* Williams and Wallace International, Inc., Toronto. (11)

Steele, T., 1981, Lamarck and immunity: a conflict resolved, *New Sci.,* 90, 360–361. (11)

Steinmetz, M., J. G. Frelinger, D. Fisher, T. Hunkapiller, D. Pereira, S. M. Weissman, H. Uehara, S. Nathenson, and L. Hood, 1981, Three cDNA clones encoding mouse transplantation antigens: homology to immuno-globulin genes, *Cell,* 24, 125–134. (10)

Stellwagen, E., and H. K. Schachman, 1962, The dissociation and reconsti-tution of aldolase, *Biochemistry,* 1, 1056–1069. (8)

Stocker, B. A. D., and P. H. Mäkelä, 1978, Genetics of the (gram-negative) bacterial surface, *Proc. R. Soc. London Ser. B,* 202, 5–30. (9)

226

Strathern, J. N., and I. Herskowitz, 1979, Assymetry and directionality in production of new cell types during clonal growth: the switching pattern of homothallic yeast, *Cell*, 17, 371–381. (10)

Strom, C. M., M. Moscona, and A. Dorfman, 1978, Amplification of DNA sequences during chicken cartilage and neural retina differentiation, *Proc. Natl. Acad. Sci. USA*, 75, 4451–4454. (11)

Strominger, J. L., H. T. Orr, P. Parham, H. L. Ploegh, D. L. Mann, H. Bilofsky, H. A. Saroff, T. T. Wu, and E. A. Kabat, 1980, An evaluation of the significance of amino acid sequence homologies in human histocompatibility antigens (HLA-A and HLA-B) with immunoglobulins and other proteins, using relatively short sequences, *Scand. J. Immunol.*, 11, 573–592. (10)

Sved, J. A., 1971, An estimate of heterosis in *Drosophila melanogaster, Gen. Res.*, 18, 97–105. (4)

Sved, J. A., 1975, Fitness of third-chromosome homozygotes in *Drosophila melanogaster, Genet. Res. Camb.*, 25, 197–200. (4)

Sved, J. A., and F. J. Ayala, 1970, A population-cage test for heterosis in *Drosophila pseudoobscura, Genetics*, 66, 97–113. (4)

Tateno, Y., and M. Nei, 1978, Goodman et al.'s method for augmenting the number of nucleotide substitutions, *J. Mol. Evol.*, 11, 67–73. (3)

Tateno, Y. and M. Nei, 1979, Augmentation algorithm: a reply to Holmquist, *J. Mol. Evol.*, 13, 167–171. (3)

Temin, H. M., 1974, On the origin of RNA tumor viruses, *Ann. Rev. Genet.*, 8, 155–178. (11)

Templeton, A. R., 1980a, Book review of *Macroevolution: Pattern and Process, Evolution*, 34, 1224–1227. (2)

Templeton, A. R., 1980b, The theory of speciation via the founder principle, *Genetics*, 94, 1011–1038. (5)

Templeton, A. R., 1981, Mechanisms of speciation—a population genetic approach, *Ann. Rev. Ecol. Syst.*, 12, 23–48. (2, 7)

Templeton, A. R., 1982, The genetic architecture of speciation, in *Mechanisms of Speciation*, M. J. D. White and C. Barigozzi (eds.), Plenum Press, New York. (2, 7)

Templeton, A. R., and L. V. Giddings, 1981, Letter to the editor, *Science*, 211, 770–771. (2)

Tomich, P. K., F. Y. An, and D. B. Clewell, 1978, A transposon (Tn *917*) in *Streptococcus faecalis* that exhibits enhanced transposition during induction of drug resistance, *Cold Spring Harbor Symp. Quant. Biol.*, 43, 1217–1221. (11)

Tracey, M. L., and F. J. Ayala, 1974, Genetic load in natural populations: is it compatible with the hypothesis that many polymorphisms are maintained by natural selection?, *Genetics*, 77, 569–589. (4)

Trippa, G., A. Loverre, and A. Catamo, 1976, Thermostability studies for investigating non-electrophoretic polymorphic alleles in *Drosophila melanogaster, Nature*, 260, 42–43. (4)

227

Tuan, D., P. A. Biro, J. K. de Riel, H. Lazarus, and B. G. Forget, 1979, Restriction endonuclease mapping of the human $^A\gamma$ globin gene, *Nucleic Acids Res.*, 6, 2519–2544. (4)

Upholt, W. B., 1977, Estimation of DNA sequence divergence from comparison of restriction endonuclease digests, *Nucleic Acids Res.*, 4, 1257–1265. (3)

Upholt, W. B., and I. B. Dawid, 1977, Mapping of mitochondrial DNA of individual sheep and goats: rapid evolution in the D loop region, *Cell,* 11, 571–574. (4)

Uy, R., and F. Wold, 1977, Posttranslational covalent modification of proteins, *Science,* 198, 890–896. (8)

Vawter, A. T., R. Rosenblatt, and G. C. Gorman, 1980, Genetic divergence among fishes of the eastern Pacific and the Caribbean: support for the molecular clock, *Evolution,* 34, 705–711. (3)

Vrba, E. S., 1980, Evolution, species and fossils: how does life evolve?, *S. African J. Sci.*, 76(2), 61–84. (5)

Vuillaume, M., and A. Berkaloff, 1974, LSD treatment of *Pieris brassicae* and consequences on the progeny, *Nature,* 251, 314–315. (11)

Wade, M. J., 1977, An experimental study of group selection, *Evolution,* 31, 134–153. (5)

Waddington, C. H., 1953, Genetic assimilation of an acquired character, *Evolution,* 7, 118–126. (11)

Walker, A., 1976, Splitting times among hominoids deduced from the fossil record, in *Molecular Anthropology,* M. Goodman et al. (eds.), Plenum Press, New York, pp. 63–77. (3)

Walker, A., and P. J. Andrews, 1973, Reconstruction of the dental arcades of *Ramapithecus wickeri, Nature,* 244, 313–314. (3)

Wallace, D. G., L. R. Maxson, and A. C. Wilson, 1971, Albumin evolution in frogs: A test of the evolutionary clock hypothesis, *Proc. Natl. Acad. Sci. USA,* 68, 3127–3129. (3)

Ward, R. D., 1975, Alcohol dehydrogenase in *Drosophila melanogaster*: a quantitative character, *Genet. Res. Camb.*, 26, 81–93. (4)

Ward, R. D., and P. D. N. Hebert, 1972, Variability of alcohol dehydrogenase activity in a natural population of *Drosophila melanogaster, Nature New Biol.,* 236, 243–244. (4)

Warren, R. A. J., 1972, Lactose-utilizing mutants of *lac* deletion strains of *Escherichia coli, Can. J. Microbiol.,* 18, 1439–1444. (9)

Weatherall, D. J., and J. B. Clegg, 1979, Recent developments in the molecular genetics of human hemoglobin, *Cell,* 16, 467–479. (10)

Webster, G., and B. Goodwin, History and structure in biology, unpublished manuscript. (5)

Weigert, M., R. Perry, D. Kelley, T. Hunkapiller, J. Schilling, and L. Hood, 1980, The joining of V and J gene segments creates antibody diversity, *Nature,* 283, 497–499. (10)

228

Weisenfeld, S. L., 1967, Sickle-cell trait in human biological and cultural evolution, *Science,* 157, 1134–1140. (2)

White, M. J. D., 1978, *Modes of Speciation,* W. H. Freeman, San Francisco. (1, 5, 7)

Williams, G. C., 1966, *Adaptation and Natural Selection,* Princeton University Press, Princeton, New Jersey. (5)

Williams, R. O., J. R. Young, and P. A. O. Majiwa, 1979, Genomic rearrangements correlated with antigenic variation in *Trypanosoma brucei, Nature,* 282, 847–849. (11)

Williamson, P. G., 1981, Palaeontological documentation of speciation in Cenozoic molluscs from Turkana Basin, *Nature,* 293, 437–443. (5)

Wills, C., and H. Jörnvall, 1979a, The two major isozymes of yeast alcohol dehydrogenase, *Eur. J. Biochem.,* 99, 323–331. (8)

Wills, C., and H. Jörnvall, 1979b, Amino acid substitutions in two functional mutants of yeast alcohol dehydrogenase, *Nature,* 279, 734–736. (8)

Wilson, A. C., 1975, Evolutionary importance of gene regulation, *Stadler Symp. Univ. Missouri,* 7, 117–133. (1)

Wilson, A. C., G. L. Bush, S. M. Case, and M. C. King, 1975, Social structuring of mammalian populations and rate of chromosomal evolution, *Proc. Natl. Acad. Sci. USA,* 72, 5061–5065. (1, 5, 7)

Wilson, A. C., S. S. Carlson, and T. J. White, 1977, Biochemical evolution, *Ann. Rev. Biochem.,* 46, 573–639. (3, 11)

Wilson, A. C., L. R. Maxson, and V. M. Sarich, 1974a, Two types of molecular evolution: evidence from studies of interspecific hybridization, *Proc. Natl. Acad. Sci. USA,* 71, 2843–2847. (3)

Wilson, A. C., V. Sarich, and L. Maxson, 1974b, The importance of gene arrangement in evolution: evidence from studies on rates of chromosomal, protein and anatomical evolution, *Proc. Natl. Acad. Sci. USA,* 71, 3028–3030. (1)

Wilton, A. N., and J. A. Sved, 1979, X-chromosomal heterosis in *Drosophila melanogaster, Genet. Res. Camb.,* 34, 303–315. (4)

Wimsatt, W. C., 1980, Reductionistic research strategies and their biases in the units of selection controversy, in *Scientific Discovery: Case Studies,* T. Nickles (ed.), Kluwer Boston, Inc., Hingham, Massachusetts, pp. 213–259. (5)

Woese, C. R., 1981, Archaebacteria, *Sci. Amer.,* 244(6), 98–122. (3)

Woese, C. R., and G. E. Fox, 1977, Phylogenetic structure of the procaryotic domain: the primary kingdoms, *Proc. Natl. Acad. Sci. USA,* 74, 5088–5090. (3, 9)

Woese, C. R., L. J. Magrum, R. Gupta, D. A. Stahl, J. Kop, N. Crawford, J. Brosius, R. Gutell, J. J. Hogan, and H. F. Noller, 1980a, Secondary structure model for bacterial 16S ribosomal RNA: phylogenetic, enzymatic and chemical evidence, *Nucleic Acids Res.,* 8, 2275–2293. (3)

Woese, C. R., J. Maniloff, and L. B. Zablen, 1980b, Phylogenetic analysis of the mycoplasmas, *Proc. Natl. Acad. Sci. USA,* 77, 494–498. (3)

Wright, C. S., 1972, Comparison of the active-site stereochemistry and substrate conformation in α-chymotrypsin and subtilisin BPN', *J. Mol. Biol.,* 67, 151–163. (8)

229

Wright, S., 1932, The roles of mutation, inbreeding, crossbreeding, and selection in evolution, *Proc. 6th Internatl. Congr. Genet.*, 1, 356–366. (2)

Wright, S., 1940, Breeding structure of populations in relation to speciation, *Amer. Nat.*, 74, 232–248. (1)

Wright, S., 1941, On the probability of fixation of reciprocal translocations, *Amer. Nat.*, 75, 513–522. (7)

Wright, S., 1960, Physiological genetics, ecology of populations, and natural selection, in *Evolution after Darwin,* S. Tax (ed.), vol. I, *The Evolution of Life,* University of Chicago Press, Chicago, pp. 429–475. (2)

Wright, S., 1968–1978, *Evolution and the Genetics of Populations,* University of Chicago Press, Chicago, vols. 1–4 (5), vol. 4 (2, 4).

Wright, S., 1982, Character change, speciation and the higher taxa, *Evolution,* in press. (5)

Wu, T. T., E. C. C. Lin, and S. Tanaka, 1968, Mutants of *Aerobacter aerogenes* capable of utilizing xylitol as a novel carbon source, *J. Bacteriol.*, 96, 447–456. (9)

Wyles, J. S., and G. C. Gorman, 1980, The albumin immunological and Nei electrophoretic distance correlation: a calibration for the saurian genus *Anolis* (Iguanidae), *Copeia,* 1, 66–71. (3)

Yanofsky, C., and M. van Cleemput, 1982, Nucleotide sequence of *trpE* of *Salmonella typhimurium* and its homology with the corresponding sequence of *Escherichia coli,* in preparation. (9)

Yanofsky, C., V. Horn, M. Bonner, and S. Stasiowski, 1971, Polarity and enzyme functions in mutants of the first three genes of the tryptophan operon of *Escherichia coli, Genetics,* 69, 409–433. (9)

Yanofsky, C., T. Platt, I. P. Crawford, B. P. Nichols, G. E. Christie, H. Horowitz, M. van Cleemput, and A. M. Wu, 1981, The complete nucleotide sequence of the tryptophan operon of *Escherichia coli, Nucl. Acids Res.*, in press. (9)

Zalkin, H., 1973, Anthranilate synthetase, *Adv. in Enzymol.*, 38, 1–39. (9)

Zalkin, H., and C. Yanofsky, 1982, Yeast TRP5: structure, function, regulation, *J. Biol. Chem.*, forthcoming. (9)

Zambryski, P., M. Holsters, K. Kruger, A. Depicker, J. Schell, M. Van Montagu, and H. M. Goodman, 1980, Tumor DNA structure in plant cells transformed by *A. tumefaciens, Science,* 209, 1385–1391. (10)

Zieg, J., M. Hilmen, and M. Simon, 1978, Regulation of gene expression by site-specific inversion, *Cell,* 15, 237–244. (10)

Zimmering, S., L. Saunders, and B. Nicoletti, 1970, Mechanisms of meiotic drive, *Ann. Rev. Genet.*, 4, 409. (11)

Zuckerkandl, E., 1976, Evolutionary processes and evolutionary noise at the molecular level. II. A selectionist model for random fixations in proteins, *J. Mol. Evol.*, 7, 269–311. (3)

Zuckerkandl, E., and L. Pauling, 1962, Molecular disease, evolution, and genic heterogeneity, in *Horizons in Biochemistry,* M. Kasha and B. Pullman (eds.), Academic Press, New York, pp. 189–225. (3)

230

Zuckerkandl, E., and L. Pauling, 1965, Evolutionary divergence and convergence in proteins, in *Evolving Genes and Proteins,* V. Bryson and H. J. Vogel (eds.), Academic Press, New York, pp. 97–166. (3)

INDEX

This book was set in VIP Century Schoolbook at DEKR Corporation. Design and production was coordinated by Joseph J. Vesely, and Fredric J. Schoenborn created the artwork. It was manufactured by the Murray Printing Company.